D1242109

GENDER
IN
URBAN
RESEARCH

URBAN AFFAIRS ANNUAL REVIEW

A semiannual series of reference volumes discussing programs, policies, and current developments in all areas of concern to urban specialists.

The **Urban Affairs Annual Review** presents original theoretical, normative, and empirical work on urban issues and problems in volumes published on an annual or bi-annual basis. The objective is to encourage critical thinking and effective practice by bringing together interdisciplinary perspectives on a common urban theme. Research that links theoretical, empirical, and policy approaches and that draws on comparative analyses offers the most promise for bridging disciplinary boundaries and contributing to these broad objectives. With the help of an international advisory board, the editors will invite and review proposals for **Urban Affairs Annual Review** volumes that incorporate these objectives. The aim is to ensure that the **Urban Affairs Annual Review** remains in the forefront of urban research and practice by providing thoughtful, timely analyses of cross-cutting issues for an audience of scholars, students, and practitioners working on urban concerns throughout the world.

RECENT VOLUMES

GENDER
IN
URBAN
RESEARCH
♦

editedby

JUDITH A. GARBER
ROBYNE S. TURNER

**URBAN
AFFAIRS
ANNUAL
REVIEW
42**

SAGE Publications
International Educational and Professional Publisher
Thousand Oaks   London   New Delhi

For information address:

SAGE Publications, Inc.
2455 Teller Road
Thousand Oaks, California 91320

SAGE Publications Ltd.
6 Bonhill Street
London EC2A 4PU
United Kingdom

SAGE Publications India Pvt. Ltd.
M-32 Market
Greater Kailash I
New Delhi 110 048 India

Printed in the United States of America

Library of Congress Cataloging-in-Publication Data

ISBN 0083-4688

ISBN 0-8039-5724-6 (cloth)

ISBN 0-8039-5725-4 (paper)

95 96 97 98 99 10 9 8 7 6 5 4 3 2 1

Sage Production Editor: Diana E. Axelsen

Contents

Acknowledgments

The idea for this volume originated with Susan Clarke, who observed with regret at the 1992 meeting of the Urban Affairs Association that work on gender was essentially absent from UAA conferences. For the 1993 UAA meeting in Indianapolis, four panels on gender issues were organized, and earlier versions of most of the chapters here were presented at that conference. We would like to thank the conference program chairs, Roger Caves and Mary Ellen Mazey, for facilitating the success of those panels. Margaret Wilder, Genie Stowers, Toni Travis, and Marsha Ritzdorf also participated on those panels, and they have contributed to this book either through their perceptive comments on the papers or their own research on gender issues. We greatly appreciate their work. The **Urban Affairs Annual Review** series editors, Susan Clarke and Dennis Judd, commented scrupulously on all of the contributions. Their readings of the chapters and the introduction provided direction to the volume and helped clarify our thinking; their moral support was vital to neophyte editors. Sage editor Carrie Mullen has been enthusiastic about this project since before the conference papers were presented, and she shepherded the book from concept to completed manuscript patiently and expeditiously.

Introduction

JUDITH A. GARBER
ROBYNE S. TURNER

■ The Gendered City

This volume identifies and examines gendered relationships that are embedded in urban life and present in the variety of localities that Americans and Canadians call community. In cities, where group membership is defined, the economy organized, democratic politics regulated, social interactions mediated, culture produced, and space exploited, how these take place has profoundly different meanings for men and women. The activities of the city too often ratify familiar gender scenarios but may also provide opportunities to resist them.

■ An Institutional Approach

Gender is a relationship between women and men. Like any relationship, gender is governed by rules appropriate to the settings in which it is reenacted and re-created. These rules structure the roles that people inhabit and assign chances for activity, movement, empowerment, and expression according to one's sex. The asymmetry of gender is the essence of patriarchy, which has three dimensions: Behaviorally, men expect to be able to bring resource advantages (such as wealth and strength) to their interactions with women. Structurally, maleness is a permanent, transcendent position that benefits all men (to different extents), no matter what individuals actually intend or desire. Ideologically, these behavioral and structural aspects are justified by a comprehensive belief system about male privilege. Unveiling and critiquing patriarchal social arrangements is what drives feminist analyses, and it distinguishes feminism from other approaches that talk about women or identify sex differences but do not seek to place their findings into a context of systematic dominant-subordinate relationships, whether historical or current.

If relationships between women and men were played out exclusively on personal terms, gender would be only a behavioral phenomenon and not the most pressing matter for urban scholars to pursue. Through its social, political, and economic features, however, gender wields tremendous institutional force—patterned authoritative decisions and procedures reflect and uphold the three dimensions of patriarchy and provide the societal backbone for relationships between individuals. Institutionalized gender also works by habituating us to the definitions of gender that are presented as possible and desirable by teachers, social workers, police, and so on. City governments construct gendered institutions directly and by administering higher-level state decisions, in addition to sanctioning gendered dimensions of nonstate institutions of religion, recreation, business, and family.

Gender relations in the city are also made more or less feasible by incentive structures that shape individual and corporate behavior. Women's dependency is effectively institutionalized by (a) welfare, job training, housing, and child care policies that label as pathological women who do not rely on men for material support or social and self worth; (b) public safety services, public transportation, and streetscapes that render women physically vulnerable; and (c) economic development strategies that produce jobs of marginal benefit to the majority of female heads of households. Hence the local state induces, even though it might not force, gender compliance.

Because gender relationships are not solely individual, the political and environmental manifestations of gender are not necessarily conscious or uncontradictory. This subtlety of relationships should be construed as evidence that they are woven into the fabric of the city and simply taken for granted, not proof that they are nonexistent. Nevertheless, the institutional elements of gender relationships may be difficult to discern; we therefore must pay very close attention both to urban institutions and to women's position in the city, relative to men's. An institutional approach will tell us about gender, which is one theme of this book, but it also provides substantive insights into the second theme, urban matters such as land use, services, housing, social control, transportation, and economic development, which urbanists currently understand mostly in terms of class, race, and individuals.

The layering of gender on top of urban issues manifests itself in the organization of production and consumption; public policies about domestic life, especially those affecting children and mothers; policing public and private spaces; and regulating land for segregation, the control of household arrangements, and growth. All are specifically urban and inescapably gendered. Furthermore, both the practice of citizenship and the practice of reducing the effectiveness of other people's citizenship through the control of local boundaries, electoral manipulation, and the imposition of dominant community norms have always been integral parts of the urban domain in the United States and Canada. In this way, collective life is determined politically, by a range of state and nonstate factors.

Urbanization reflects the workings of large processes and ideologies besides patriarchy—notable are capitalism, modernization, and colonialism, on a global level (Van Vliet, 1988; Wilson, 1991, pp. 121-134), and capitalism, development, and racism, on a national level. Women are not affected equivalently by the institutional manifestations of these forces, nor do women participate equally in them. In North America, most women view cities not through the eyes of abstract individuals but from their place on "multiple hierarchies" (Crenshaw, 1993, p. 114) of sex, nationality, race, sexual preference, and class that are constructed in distinctly urban terms. At the intersection of the hierarchies, where women are doubly or triply burdened, the nature of gender becomes conflicted and muddled. In any urban setting, gender does not act in isolation from other social relations or leave one, unique impression on women's urban experiences (Rose, 1993, chap. 6; Sandercock & Forsyth, 1992).

■ Gendered Analysis and Feminist Theory

The exploration of gender in this book draws on two general avenues for studying urban gender issues, each of which is compatible with an institutional perspective. One option is employing gender, rather than class or race, as a "category of analysis" (or as a "topic of inquiry" for descriptive purposes) (Silverberg, 1990, 1992). Here, urban conditions are viewed in terms of gender's interactions with state and social structures, where gender acts as a cause, effect, or premise of those structures. Although, conceivably, one could conduct gendered analyses without having in mind an agenda for undoing patriarchy, the focus on gender (as an unequal relationship) does presume acceptance of a major tenet of feminism. A second option is feminist theory, whose critical stance (Hennessey, 1993, pp. 12-13) is prerequisite to the project of reconceptualizing gender relations. The prescriptive features of feminist theories are therefore usually explicit and integral, although feminist theories diverge fundamentally over whether gender *inequalities* can or should be viewed as mere gender *differences* and what order of social change is required to achieve feminist goals. In neither option are the frameworks or women themselves treated as mere subsets or "cases" of the normative and analytical paradigms common in North American urban research, such as the growth machine, regime theory, Marxism, pluralism, postmodernism, and public choice. Although this volume does not uniformly reject any urban theory, neither is it uncritical of the secondary position or nonposition accorded gender in the "questions, categories, and models" (Silverberg, 1992, p. 370) applied to urban problems.

Nongendered frameworks may be incorporated in formulating descriptions, analyses, and prescriptions about gender issues in cities; nevertheless, the importance of such research lies in revising both the findings and the paradigms of urban research. Gendered analyses and feminist theory contribute responses to what Silverberg (1992) calls "common feminist . . . concerns" (p. 375) articulated within the general field of urban studies and outside of it. Keeping in mind that attempts to devise a unitary feminist agenda are in some sense inimical to feminism (Sandercock & Forsyth, 1992), both research options do provide room for claims about previously ignored facets of urban existence—those related to women's shared experiences and fortunes, as well as their conflicting identities. This effort reflects the postmodern critique of culture and society that has attracted feminists and urbanists, but it seeks to ensure that our skepticism about objective knowledge is not so "relentless" that power is left as "the sole determining factor" in assigning authority to observations about locality and space (Bondi & Domosh, 1992, p. 210). Gendered analyses and feminist theories permit the existing order of power and privilege—and therefore the institutions that govern relationships in urban settings—to be challenged directly on the basis of gender inequalities.

■ Public and Private Spheres

Feminist challenges of this sort are overwhelmingly animated by a desire to reintegrate the public-private distinction that underlies the design and governance of North American cities and some considerable subset of urban gender relations. Feminist theorists from widely disparate standpoints balk at classical liberalism's confidence that the public and private elements of social life can be defined and that it is natural for them to be kept apart. Articulating geographical, economic, or political linkages between public and private gender relations is a goal, stated or implicit, of most feminist work on cities. Because the tenets of the liberal dichotomy do not apply equally to women who are not White, heterosexual, and middle-class, gendered urban analyses must pay particular attention to the existence of dissimilar kinds of public and private spheres.

Gender succeeds as a mode of social organization and an exercise of power because institutions (economy, politics, law, the academy) that are typically of the male domain are removed from those (home, family, and culture) whose cultivation is assigned to women. Some separation of public and private is necessary to the maintenance of gender roles for two reasons. First, because the spheres are distinct, one may be set above the other—public affairs are more serious than private and a more proper subject of political discourse and public resources. For this reason, jobs and housing occupy city governments, but not domestic labor or the concept of the home. Second, the private sphere is a haven where gender roles are perpetuated by women and where men's ability to carry out public activities (including "private" economic endeavors) is facilitated. Women's supposed predominance here can be usurped—and women dominated—because the sanctity of the nuclear family is legally respected. Hence police and local officials do not often treat prostitution in terms of children who flee abusive homes to the streets or the sexual exploitation of women but, rather, as a threat to "normal" neighborhoods that can simply be shifted to deserted zones.

A public-private dichotomy of this sort coincides neatly with the preeminent feature of North American cities, the division of space into uses that are public and common versus private and owned, or cities versus suburbs (Saegert, 1980). The correspondence between gender roles based on separate spheres and the urban function of regulating land is not incidental but actually points out that gender roles are institutionalized at the local level and suggests how routinely this happens. Without much acknowledgment or controversy, the organization of cities buttresses personal gender relationships in which women are primarily at home and men primarily at work.

This formal sexual division of social labor is obviously only a rough approximation of the urban world, and the chapters in this volume demonstrate that a rigid application of the public-private distinction in gender analyses is useless. The vast majority of Black and immigrant women in cities have a long tradition of paid work, as do working-class and single women, generally. Women of color have occupied the public sphere, but at the lowest strata, and their role in social reproduction has been appropriated to support the gender relations of White, suburban families as maids and nannies. Their own private gender relations have been subject to state control (see Rose, 1993, pp. 125-127), even if they are not state clients. In the late 20th century, gender has also been rendered less predictable by the value changes and policy gains of the feminist movement; economic restructuring has entailed role reversals, as men become unemployed and women increasingly enter the workforce; and fewer urban households include adult men, by choice or circumstance. In the midst of gender changes, the public and private spheres of the city are increasingly contradictory and difficult to negotiate. For many inner-city females, the contradiction is disastrous: Work and home collide as economic and territorial transformations in metropolitan areas demand more time but do not guarantee the same standard of living.

Finally, it should not be overlooked that women often do not observe gender roles. In politics and society, the quiet violation and open confrontation of gender conventions influences the public and private divisions of the city. In urban scholarship, both the fact of separate, gendered spheres and their instability are too important to ignore. Nevertheless, the traditional gender role division indeed remains an assumption of most public policy and academic research.

■ Rethinking the Paradigms of Urban Research

A focus on gender in the city calls into question whether it is sufficient for urban researchers to focus on the *economic* logic behind the separation and obfuscation of public and private spheres. Urban scholars are greatly unaccustomed to thinking about gender, let alone uncovering cities as sites of institutionalized patriarchy. Of course, it is commonly recognized by now that women are afraid to venture into city streets or parks, that U.S. and even Canadian central cities are increasingly defined by large numbers of poor female-headed households, and that female electoral success is most attainable at the local level. In the overall picture of the gendered city, however, the true pervasiveness of gender can be represented only by disclosing and ordering the various snapshots of women's place in patriarchal institutions, because disparate events may appear to be random, and by explaining and theorizing the mechanisms by which such gender differences come to dominate the city, because they may have no apparent context or origin. Given these tasks, the research agenda properly includes descriptive and analytical, as well as normative elements.

A necessary point of departure in bringing together gender and urban research is to recognize that gender relations are not natural features of cities but are social constructs like any other institutionalized urban relation. Gender is often falsely treated as natural, even more so than are the racial and class biases of cities. If they are not passed over altogether, urban gender differences and sexual inequality may be explained away as the unpatterned consequences of individual choices, subsumed under race and class structures, or used to show how men are at risk. Yet gender and women's disadvantages are pervasive and institutionalized dynamics of the construction of urban places and should be treated as such in the analysis of cities.

As an antidote to the dominant view of gender in cities, three broad questions are addressed throughout this book and tentative answers offered: How is gender institutionalized in distinctively urban and local ways? Once ensconced in institutions, how does gender physically shape the city, define men's privileges over women, and differentiate between women? How do women recognize the gendered rules that compose urban institutions and attempt to construct their own, preferred versions of proper gender relations and institutions? Any of these questions can be approached empirically or normatively, depending on whether we want to identify the mechanics of gender at the local level (including the interests in maintaining and undermining gender relations) or develop a coherent set of feminist values about cities or for use in cities.

This volume springs from the belief that institutionalized gender relationships are an intimate part of what the terms *urban* and *local* mean in the United States and Canada. The idea of gendered institutions can provide a theme for and be incorporated in research perspectives across social science disciplines. Like urban studies, women's studies is often conceived of in an interdisciplinary fashion; following this principle, this book intentionally spans several fields—the authors teach Africana studies, criminology, geography, planning and urban studies, political science, and sociology. Previously, gendered research on cities has defined urban most frequently in spatial and design terms—this is the privilege of the geographers, planners, and architects who have been working on gender issues longer, and with more theoretical success, than those in disciplines with nonenvironmental orientations. Feminist discussions of urban politics have lagged behind in a particularly pronounced fashion; for this reason, politics or public policy are treated in some form in each of the chapters.

Methodologies and Theories

Feminist scholars themselves differ about how research should be conducted and on what grounds. In response to the lack of a cohesive feminist methodology or theory, feminist urbanists have argued for the development of a comprehensive, encompassing field of research that recognizes women as a social group in urban structures (Masson, 1984), and some recent research presents a fairly unified theoretical framework to account for gender relations in cities. However, these efforts do not fully resolve the tendency of gendered urban research to be derivative of other theories (Bondi, 1991), because they take mostly an economic-structural view of urban phenomena and do not leave much room for interpretations other than that social reproduction is the most fundamentally gendered aspect of the local state (e.g., Fincher, 1989; Gamarnikov, 1978; Masson, 1984). Feminist social science must also guard against imposing a hegemonic, "scientized" scholarship on gender research. Such an approach purports to exclude researchers' values from the selection of a research agenda and the conduct of analyses and may devalue the pertinence of women's personal experiences to feminist work (McClure, 1992).

Certainly, we need both complex gendered interpretations of the capitalist forces that mold cities and systematic studies of the facts of women's urban lives. Gender studies rooted in initially nongendered frameworks may, however, not be able to move beyond the categories of previous theories. The tendency to stop short of a genuine gendered analysis in urban inquiries may be resolved by conscious attempts not to "accord the gender component . . . an inferior conceptual status" (Rose, 1989, p. 120) or to presume that it never mediates other variables.

A variety of methods and theories are appropriate in approaching the gendered nature of urban institutions and places, especially because the field is still small and emerging, relative to the reigning political economy and sociological approaches. The pluralism of the present volume reflects these uncertainties, and the contributions do not come to a collective conclusion about how to resolve these deep differences. Theoretically, the chapters place different value on the (potential) contribution of postmodernism, feminist theories, and structural political economy to making sense of the gendered conditions of cities. Analytically, whereas certain authors focus on public discourse, others identify gender relations through political behavior, state actions, social roles, history, or spatial forms. Methodologically, case studies, surveys, personal interviews, and aggregate data are used to portray the characteristics of women's roles in the communities, politics, and environments of urban areas.

■ Speaking Urban to Gender

Although equally as significant as the gender focus, the urban focus of these works is not always present in feminist examinations of phenomena that take place in cities. Taking urban factors seriously on their own terms, cities are not simply the scenery for the playing out of gender; place is central, not incidental, because urbanization and locational differences construct social relations such as those between women and men. The forces encompassed within the general notion of urbanization, regarding population concentration, social and cultural heterogeneity, intensity of land use, economic and technological development, and sex roles, are present in all social structures of cities. Just as urban political economists have discovered that numerous locationally specific factors exert independent influence over the particular imprint that urbanization makes on any given place, locality is an essential component of our understanding of the gendered dimensions of urban institutions. For urban researchers, local variations do not impede the development of theoretical or analytical insights about national or global social structures but are an integral part of these efforts. Indeed, because feminist frameworks are often alone in acknowledging that people are affected by multidimensional and perhaps conflicting social structures, they are especially well suited to investigating cities, if gender is not abstracted from the overall urban context or the particular urban setting.

Bringing the findings of urban research to gendered analyses and feminist theories has encouraged forays into new research ground. In the chapters of this volume, a wide range of urban topics crossing disciplinary boundaries are studied. In some cases (downtown development, economic restructuring, elections, race and class stratification, public housing, policy making), the general issues that are examined in this book have previously been approached by urbanists but from perspectives that rarely include or tolerate gender analyses. In others (styles of representation, rape and domestic violence services, welfare reform, citizenship, representation, political theories of community), the topics have sometimes been studied as women's and feminist issues but not typically as urban issues. For each of these topics, the connection is possible only because a gendered or feminist standpoint precedes the research, but the urban dimension is vital.

Canada and the United States

Despite evident comparisons between American and Canadian cities and local idiosyncrasies, national political and social attributes mold local institutions, gender relations, and the quality of women's urban lives (see Andrew & Moore Milroy, 1988, pp. 3-6). In spatial terms, U.S. urban areas are more decentralized and suburbanized, whereas a higher proportion of Canadians live in central cities or under metropolitan governments; relatedly, American urban mass transit is less advanced. In Canada, the gender, class, and household composition of gentrifiers have been more diverse, with less disruption to existing neighborhoods (Rose, 1989). There is notable racial and ethnic concentration in Canada's largest cities, Halifax, and the urban centers of the west with large aboriginal populations, but segregation is a far more pervasive, constitutive feature of U.S. cities. Abject poverty, violence, and social discord dog American urban life, whereas in Canada these are mitigated by somewhat more generous social programs funded from above. Politically, U.S. local governments have more legal powers and autonomy; they are therefore freer both to respond to advocates for democratic change *and* to pressures against it. Restructuring and reduced military spending have cut massive swaths across the U.S. Sunbelt and Rustbelt; in Canada, industrial decline (in Quebec and southern Ontario), postindustrial growth (in Vancouver and Toronto), and the death of Atlantic fishing have wrought more confined urban changes. Thus, although patriarchy is embedded within Canadian urban systems as it is in the United States, Canadian women face it under a different set of circumstances—characteristic spatial, social, political, and economic structures. This variation produces distinct gender biases and affects women's ability to minimize their impacts or successfully challenge them.

■ Terms of Reference

The 15 chapters that follow uncover layers of institutional structures, processes, and outcomes that shape urban gender relations and serve as the topics for the authors' analyses of urban phenomena. Whether the venue is a house, neighborhood, automobile, workplace, shelter, campus, or council chambers, the authors identify gendered relations operating within the institutions of city life. We categorize urban gender relations according to three terms of reference: *community*, *politics*, and *environments*. The terms make sense separately, but they are related in that all are granted form and legitimacy by governments or supported by state-market collaboration. In the three Parts of the book that correspond to these terms, both grassroots activity and public policy are explored.

The chapters that make up Part 1 consider some basic questions about the relationship between city and community for women and women's identity as urban citizens. How community and citizenship are given meaning affects women's capacity to make urban polities responsive to their interests in the city. Staeheli and Clarke initiate this discussion by critiquing the restrictive, state-oriented grounds of citizenship set out in liberalism and the notions of class used by urban political economy that exclude gender and race. Neither framework is able to address the transformation of women's political activities in the face of urban economic restructuring. They suggest a new theoretical understanding of the roles of women as actors within the institutional framework of the local state and recast our thinking about the limits and opportunities for women as citizens in a patriarchal locus.

The parameters of women's citizenship are set in part by the conception of community that operates in cities. Garber discusses competing political theories of community and their implications for feminist ideals. She stresses, as do Staeheli and Clarke, that the linkage between place and politics is vital for women; however, traditional community in the United States and Canada marginalizes women and other outsider groups. Because many women value local community, community organized around democratic political purposes in particular places is a better alternative than the identity groups proposed by some feminists.

Appleton shifts the discussion of community to an exploration of the dynamics of public and private local "gender regimes." She contrasts the more open social and economic roles permitted heterosexual women, gays, and lesbians in cities with the rigid gender roles prescribed by the nuclear family in suburbs. The dominance of a particular patriarchal system determines how communities adapt to social, political, and economic changes over time.

It is clear that public and private spheres of activity may be blurred even within patriarchal norms, subject to the geoeconomic constraints of the modern metropolis. For example, women's entrance into the postindustrial labor force will have profound domestic and community effects. Jezierski describes efforts by women to seek economic parity amid global restructuring in Pittsburgh and Boston. Some women have lost economic status, whereas others have opportunities to extend their economic status through adjoining institutional outlets in political office, community groups, and business groups. Workplace identity may be augmented by community identity as women challenge the gender dimensions of urban political economy.

Where Appleton and Jezierski outline the mixed results of going outside existing patriarchal institutions, Rabrenovic outlines the possibilities for change provided by women's collective action in neighborhood groups in Schenectady, New York, and Chelsea, Massachusetts. These low-income White and Latina women abandoned traditional gender expectations within their ethnic and economic communities and the formidable institutions of police and schools. It is evident from Rabrenovic's account of these two groups organized and directed by women that there are obstacles to resisting the institutionally derived notion of community standards, but women's own focus may prevail.

These chapters explore citizenship and community at the juncture between normative feminist theory and urban theory. Whether in feminist-organized communities, inner cities, ethnic enclaves, or suburbia, the struggles presented demonstrate the difficulty faced by women within existing institutional orders such as patriarchy and capitalism. The application of these theories to urban problems sheds light on the political processes and actions that are the subject of Part II.

A basic frame of reference for understanding community and citizenship is that actions and relationships are divided into public and private spheres that correspond to political and domestic life. However, the norms of private patriarchy carry over to politics, and it is no coincidence that women are only now making strides in election to public office and influencing policy decisions, particularly at the local level. These changes rest on the institutional processes of the public sector and the perceptions of political actors (officials and citizens), and they set the scene for gendered analyses of politics.

Andrew begins with an important discussion of how mobilization affects the public agenda of Canadian local government. Her study of local policy planning on the issue of violence against women in Ottawa is the backdrop for exploring how political discourse is related to women's success in garnering attention, legitimacy, and policy response from policymakers. The evolution of the Canadian women's movement provides markers for understanding how institutional perceptions and responses change over time. These can be turned from inattention to feminist issues to responsibility for addressing them.

Beck explores the politics of public policy making from the perspective of city councilors' views about affordable housing in suburban New Jersey. State requirements about housing equity issues forced local elected officials to articulate their social and economic concerns. Within that context, two types of politics are at work: formal, procedural liberalism and less formal, relational politics. Female representatives often embrace relational reasoning in their public decisions, and their gravitation to that form is based on gendered life experiences. Beck's interviews suggest that women bring a specific frame of institutional reference to the public sphere.

In Richmond, Virginia electoral politics, race, and class structures are enmeshed with gender bias. Randolph and Tate show how the election of an African American woman to city council brought forth a tension with the city's traditional, patriarchal structures. Historically, both issue framing and constituency representation had been structured by race relations and class relations within the Black community. The explosive response to the challenge to this structure by a woman indicates that the local political arena is protected by complex power structures derived from public and private patriarchy, regardless of race.

If Andrew, Beck, and Randolph and Tate are correct that women's participation in politics represents a change in the institutional status quo, then MacManus and Bullock preview the future of the local public sector. Recent trends in the election of women to municipal and county office in the United States show that women as a whole are facing fewer gender barriers; minority women are making significant advances in terms of number elected to office but still have additional burdens. Local office is often a stepping-stone to higher office, and that bodes well for gender reform within the public sphere. Although, increasingly, women take the same route to local elected office as do men, MacManus and Bullock's evidence suggests that women's interests and policy goals may differ from men's once they are elected.

These chapters demonstrate how assumptions about the public sphere may change when a gendered analysis is introduced. In understanding institutionalized gender biases and patriarchal systems operating within the public sector, important commonalities with the private sphere appear. The recognition that gender bridges the spheres of the city drives Part III, which concerns how women interact with state-mediated social and spatial forms in a series of urban environments.

Urban environments are constructed around the delivery of public services, the development of policies, and the contours of built form. These shape women's ability to cope with complex urban locations, largely through the responsiveness of public and private organizations to the needs of diverse groups of women and children. Although we may lose some parsimony by subdividing urban populations into gender, class, race, and age, the premise of this section is that environments are not uniform and neither are their impacts on unlike population groups. Thus the quality of local government's responses to social problems will likely be evaluated far differently than under a generic approach.

Kathlene addresses the issue of campus rape. Her survey of students who have been assaulted reveals that, especially in the case of acquaintance rape, they view themselves from society's perspective. Efforts to seek help are conditioned by the existing institutional processes and mores, which often revictimize women. Kathlene suggests how these powerful environmental forces might be revised so that policies and programs in the university and community hear women's voices. The next two chapters suggest similar patterns of environmental barriers and misinterpretations surrounding service delivery to women but compounded by race, class, and ethnic biases.

Young and Miranne analyze the child care component of the welfare reform program Project Independence, as it is administered in New Orleans. Continuing the theme that women have a variety of help-seeking preferences, the authors point out that the success of welfare-to-work policies depends on child care service options that accommodate the situations of poor, African American urban females. The stated preferences and needs of the participants for more flexible, comprehensive child care arrangements should guide policy making and implementation.

Santiago and Morash analyze the service use patterns of Latina battered women and show, again, that women's help-seeking behaviors have various origins that must be reflected in program design and delivery. Their survey data reveal that because of the cultural and language characteristics of Latino communities, Latinas may not want or be able to use urban domestic violence services designed for Anglo and African American women. Institutional decisions on program delivery must be cognizant of those variations and develop strategies accordingly.

Beyond the realm of cultural norms and the biases of state institutions, spatial and economic contexts affect women's commuting patterns, which affect their employment opportunities. Johnston-Anumonwo, McLafferty, and Preston review the employment opportunities for African American and African Canadian women through this lens of spatial access to jobs. Spatial access, measured by the time and distance of one's commute, varies by gender and within gender, by race, class, and marital status. Transportation services must be sensitive to user needs and women's circumstances, because existing conditions such as spatial patterns structure women's access to urban environments.

For women in cities, housing is perhaps the major factor that defines their relationship to urban environments. Spain examines the social purposes behind American public housing today; its similarity to the medieval beguinage, a shelter for unmarried women, shows how the state uses its institutional power to control surplus populations, in this case poor and mostly single women. Instead of adapting the institution to a changing clientele increasingly of unwed minority mothers, public housing authorities have abandoned their support function and adopted the role of detached warehouse administrator. This leaves atypical tenants to create jobs for themselves within the project confines, which some women have done successfully.

On a larger scale, economic pressures within cities have displaced poor women from their downtown neighborhoods and constrained their life choice options by controlling the availability of services, housing and employment opportunities, and land use patterns. By investigating the planning and participatory mechanisms used within the institutional decision processes of four cities, Turner explores how cities in the Sunbelt have dealt with the changing social and economic conditions in their downtowns. She concludes that spatial opportunities for women are affected by access to the development decision process. Cities that build linkages between planners and the resident population are more likely to include the interests of women, such as neighborhood priorities, in their development plans.

■ **Gender in Urban Research**

We offer this volume not only as part of the existing body of work on women and cities but also as an introduction to urbanists who have not considered the implications of gender in their research. Gendered analysis of familiar urban topics provides new insights for our understanding of urban studies, not only by counting women who had formerly been invisible but by integrating their presence into our explanatory frameworks and normative insights.

Feminist inquiries pose alternatives to the dominant urban theory constructs that base their definitions of city systems on gendered spatial, political, and economic forms. To address urban gender relations (although perhaps also class and race relations or sexuality and ethnicity) is to accept that women have particular relationships to urban institutions, that individual women's lives as well as their collective self-direction are heavily affected by issues such as those explored by this book's authors, and that women in cities attempt to construct their own versions of gender relations and institutional responses outside existing ones. It is also to judge economic development policies, service delivery strategies, planning, and democratic processes—the most traditional local concerns—as failures if they reinforce the inhospitability of cities for a signification portion of the population.

This volume seeks to bridge feminist theories and theories of the state. The selections presented here serve as examples of how this may be done; they are not exhaustive of the issues that could be considered, such as homelessness, poverty, access to health care, AIDS, gay and lesbian urban activism, aging, and environmental conditions. Indeed, these matters should be considered, because gender discourse deserves stimulating theoretical and empirical treatments in urban scholarship. We hope this volume makes a contribution toward that end.

REFERENCES

Andrew, C., & Moore Milroy, B. (1988). Introduction. In C. Andrew & B. Moore Milroy (Eds.), *Life spaces* (pp. 1-12). Vancouver: University of British Columbia Press.

Bondi, L. (1991). Women, gender relations and the "inner city." In M. Keith & A. Rogers (Eds.), *Rhetoric and reality in the inner city* (pp. 110-126). London: Mansell.

Bondi, L., & Domosh, M. (1992). Other figures in other places: On feminism, postmodernism, and geography. *Environment and Planning D: Society and Space, 10*, 199-213.

Crenshaw, K. W. (1993). Beyond racism and misogyny: Black feminism and 2 Live Crew. In M. J. Matsuda, C. R. Lawrence III, R. Delgado, & K. W. Crenshaw (Eds.), *Words that wound: Critical race theory, assaultive speech, and the First Amendment* (pp. 111-132). Boulder, CO: Westview.

Fincher, R. (1989). Class and gender relations in the local labor market and the local state. In J. Wolch & M. Dear (Eds.), *The power of geography: How territory shapes social life* (pp. 93-117). Winchester, MA: Unwin Hyman.

Gamarnikov, E. (Ed.). (1978). Women and the city [Special issue]. *International Journal of Urban and Regional Research, 2*(3).

Hennessey, R. (1993). *Materialist feminism and the politics of discourse.* New York: Routledge.

Masson, D. (1984). Les femmes dans les structures urbaines: Aperçu d'un nouveau champ de recherche, *Canadian Journal of Political Science, 17*, 753-782.

McClure, K. (1992). The issue of foundations: Scientized politics, politicized science, and feminist critical practice. In J. Butler & J. Scott (Eds.), *Feminists theorize the political* (pp. 341-368). London: Routledge.

Rose, D. (1989). A feminist perspective of employment restructuring and gentrification: The case of Montréal. In J. Wolch & M. Dear (Eds.), *The power of geography: How territory shapes social life* (pp. 118-138). Winchester, MA: Unwin Hyman.

Rose, G. (1993). *Feminism and geography: The limits of geographical knowledge.* Minneapolis: University of Minnesota Press.

Saegert, S. (1980). Masculine cities and feminine suburbs: Polarized ideas, contradictory realities. *Signs, 5*, S93-S108.

Sandercock, L., & Forsyth, A. (1992). Feminist theory and planning theory: The epistemological linkages. *Planning theory, 7/8*, 45-49.

Silverberg, H. (1990). What happened to the feminist revolution in political science? *Western Political Quarterly, 43*, 887-903.

Silverberg, H. (1992). Gender studies and political science: The history of the "behavioralist compromise." In J. Farr & R. Seidelman (Eds.), *Discipline and history: Political science in the United States* (pp. 363-381). Ann Arbor: University of Michigan Press.

Van Vliet, W. (Ed.). (1988). *Women, housing and community.* Aldershot: Avebury.

Wilson, E. (1991). *The sphinx in the city.* London: Virago.

Part I

Community

1 Gender, Place, and Citizenship

LYNN A. STAEHELI
SUSAN E. CLARKE

In this chapter, we contend that attention to gender relations offers insights and alternatives that can enrich urban political economy views of local politics. Broadly speaking, urban political economy has been concerned with relations of subordination and domination in urban politics rooted in the class relations of capitalism. To this end, analysts have focused on the actions of elites to shape economic and political practices to further their economic interests. We argue that this focus has resulted in a body of theory that is overly economistic in its explanations and state centered in its research focus. Although it is narrowly conceptualized in its focus on class divisions and elite actions, it is broadly argued as encompassing all that is theoretically significant to urban politics. Such a position, however, ignores the very important ways in which class relations are constructed through and in conjunction with gender and race relations. Understanding the roles of gender, race, and other non-class-based social divisions within the labor force is particularly important in contemporary rounds of economic restructuring, which is segmenting the labor force in new and ever more complex ways. Further, the focus on elite actions characteristic of much urban political economy research shifts attention from the agents (e.g., women) and sites of struggle (e.g., households and communities) that may be generative of new social and political movements in response to economic restructuring. Thus the parsimony that is gained by ignoring the gendered nature of urban politics may be at the expense of salience.

In the following sections, we argue for a new, gendered conception of urban politics and demonstrate its analytical power. We first identify

AUTHORS' NOTE: This research was supported in part by a National Science Foundation Grant (SES-9211679) and the Center for Public Policy Research, University of Colorado.

some of the epistemological and analytical assumptions of urban political economy that have masked the theoretical importance of gender relations; in so doing, we suggest an alternative conceptualization in which gender relations are incorporated. To demonstrate how this conceptualization can enrich our understandings of urban political economy, we apply it to an analysis of citizenship. Much of the literature on citizenship focuses on class relations as they shape social rights, but class is only one dimension of citizenship. We demonstrate the ways that gender relations intersect with other social relations to exclude social groups from the full rights of citizenship and to shape political action. In so doing, we argue for a more catholic, more contextual approach to urban politics that incorporates a greater variety of social groups as they respond to the political and economic restructuring of contemporary cities. The argument we present is both analytical and normative. We demonstrate the usefulness of citizenship as an analytical construct to help us understand the ways in which social relations affect the standing of individuals and social groups in the polity. We also suggest ways of thinking about citizenship that will allow greater roles for women's political activities.

■ Reassessing Urban Political Economy Approaches

An appreciation of critical theory and a focus on gender relations can be used to reformulate urban political economy and give it added salience in explanations of contemporary rounds of restructuring. We use critiques raised in critical social theory to explore the epistemological and analytical assumptions that limit the usefulness of urban political theory and, indeed, all social science. We then broaden urban political theory in light of these critiques.

Critical Theories of Science

Critical social theorists have critiqued the ways in which science has been practiced and codified. Although these critiques are varied, wide-ranging, and in some cases, conflicting, there is widespread agreement that "science as usual" has not led to a sufficient understanding of the ways in which social relations shape our lives. Many of these critiques are relevant to rethinking the limitations of urban political economy. In this discussion, we focus on critiques made by feminist theorists because their epistemological arguments are connected with a substantive concern for gender relations. Thus much of the argument presented here is also made by nonfeminist theorists, and we do not claim or imply otherwise. Rather, we draw on feminist theorists to focus the discussion on the ways in which mainstream scientific practice can obscure or lead to misrepresentations of gender relations.

Of particular relevance is the argument of many feminists that mainstream, or "malestream," approaches to science assume objectivity of researchers with respect to their decisions related to problem definition, use of methods, measurement, and evaluations of research subjects. Most social science constructions assume that the influence of the values, motives, and material conditions of researchers can be held in check during the research process. Feminists and other critical theorists reject this position and argue that all scientists and scientific activity are situated. As Harding (1991) argues,

> An outpouring of recent studies in every area of the social studies of the sciences forces the recognition that all scientific knowledge is always, in every respect, socially situated. Neither knowers nor the knowledge they produce are or could be impartial, disinterested, value-neutral, Archimedean. The challenge is to articulate how it is that knowledge has a socially situated character denied to it by the conventional view. (pp. 11-12)

We must acknowledge that theories presented as universal or objective are local, historically rooted, and partial (Young, 1990). The problem of objectivity is compounded when research implicitly takes the experiences of males as neutral and universal, when they are, in truth, subjective and partial. Thus the challenge for social science is to explore the ways in which research is conditioned by its context and by dominant cultural and analytical paradigms.

Some of the most important elements of the culture reflected in analytical paradigms are relations of subordination and domination. Urban political economy has focused on these relations as they pertain to class divisions. The subordination of women, however, cannot be reduced to or subsumed under class relations. A more inclusive theory, then, should explore the ways in which gender and class relations overlap, reinforce, and to some extent are constitutive of each other. In arguing for the social construction of knowledge and its implications for how knowledge is used, Dorothy Smith (1987) argues that women are oppressed through overlapping "relations of ruling." These relations include the forms of socially constructed and accepted knowledge, organized practices and institutions, and questions of experience and agency. The relations of ruling combine to make women, and thereby gender relations, invisible and analytically unimportant in most realms—including politics, economics, and theories of them.

We proceed from the position that urban political economy reflects the gendered social relations in which it was developed such that women and gender are made invisible and theoretically inconsequential. We can identify the effects or influences of these relations of ruling in key epistemological and analytical assumptions of urban political economic theory. These assumptions include (a) a separation of public and private spheres of action, (b) an overgeneralized notion of what constitutes public action, (c) a narrow conception of political agents, and (d) a restrictive notion of coalitions. Each of these assumptions is examined in turn.

The Separation of Public and Private Spheres

Most political theories—including urban political economy approaches—conceptualize separate public and private spheres of activity. The public sphere includes economic and political relations and activity, generally at the community, state, and national levels. This is contrasted with the private sphere, which includes personal and family relations and activities, generally at the levels of the body and the household. As Marston (1990) and Pateman (1989) argue, this separation has been a central assumption of analyses of the state and politics. This duality, however, is problematical for those who wish to broaden political economy approaches to include gender relations.

Dualisms, and binary thinking generally, inscribe processes of domination and discrimination into language and theory. This is because the terms in dualisms

> gain their meaning only in *relation* to their difference from their opposi-
> tional counterparts. . . . Difference is not complementary in that the halves
> of the dichotomy do not enhance each other. Rather the dichotomous
> halves are different and inherently opposed to one another. . . . Such
> dualities rarely represent different but equal relationships. . . . Thus whites
> rule Blacks, males dominate females, reason is touted as superior to
> emotion in ascertaining truth, facts supersede opinion in evaluating knowl-
> edge and subjects rule objects. (Collins, 1986, p. S20)

Thus, in dualistic thinking, the leading term *public* is accorded primacy, and *private* is derivative.

This has important analytical implications. In any piece of research, one strives to be sure that the "big," "important," and "causative" forces are incorporated. Expediency and/or parsimony, however, may lead one to ignore or delay analyses of "small," "subsidiary," and "derivative" issues. To the extent that dualistic language and thinking devalues the private sphere, and the issues and actors associated with it, ignoring the private sphere is acceptable and understandable from analytical and theoretical perspectives.

The public/private dualism and the resulting devaluation of the private sphere has ramifications for the study of women and gender relations. Put baldly, the public sphere is associated with the domains of production, paid employment, politics, and men, whereas the private sphere is seen as the site of reproduction, families, informal and/or unpaid employment, and women (Vaiou, 1992). These dualistic constructs contribute to a partial view of urban political economy because they obscure the extent to which the public sphere is constituted by women's activities and relations in the household and private sphere. Similarly, these dualisms also obscure the extent to which the private sphere may be transformed by relations and actions in the public sphere. From a normative perspective, Gould (1988) argues that the separation of economic, political, and social relations across public and private spheres inherently limits our ability to conceive of and act to achieve a true equality.

Overgeneralization of the State

The problems wrought through the dualistic construction of public and private spheres are compounded by a tendency to overgeneralize the state almost to the point of equating it with the public sphere. In political economy approaches, there is generally an assumption of a single public sphere that is limited to the state in the operationalization of research. Political activity is narrowly defined in terms of engagement with the state. Other actions and relations connected with the market and non-state-centered discourse are outside the realm of politics (Fraser, 1990; Nelson, 1984). Thus the range of issues, actions, and relations open to analysis is further narrowed.

This narrowness, borne of dualistic thinking and overgeneralization of the state, is one reason many urban analysts look beyond urban political economy theories to explain contemporary events. These analysts demonstrate the central role of the household in mediating and transforming economic restructuring; indeed, they argue that households are the microstructures through which global economic changes are experienced and enacted. As examples, Lawson's (1992) work demonstrates the way in which global economic conditions have changed the employment relations in the workplace and in the home. Gibson and Graham (1992) argue that one result of restructuring is to make households one of the primary sites of class struggle. Our own research examines the forging of a new politics of work and community in which households are repositioning themselves with respect to the state (Clarke & Staeheli, 1993). We argue that one could not account for these new trends through an examination of the state separated from other "public" activities (such as employment) and from the private sphere.

These works underscore the need to reconstruct Habermas's (1989) notion of a public sphere to contest the boundaries of public and private concerns and the arenas in which they should be addressed. By implication, the analysis shifts from a state-centered view. Fraser (1990) and Young (1990) argue for a redefinition of public that recognizes heterogeneity and the reality of social group differentiation. Young links her redefinition to a reversal of the public/private dualism. Rather than assigning everything that is not public to the private sphere, she defines the private sphere first as "that aspect of his or her life activity that any person has a right to exclude others from" (p. 119). Thus actions and relations are not a priori excluded from the public sphere:

> This manner of formulating the concepts of public and private . . . does not deny their distinction. It does deny, however, a social division between public and private spheres, each with different kinds of institutions, activities, and human attributes. The concept of a heterogeneous public implies two political principles: (a) no persons, actions, or aspects of a person's life should be forced into privacy; and (b) no social institutions or practices should be excluded a priori from being a proper subject for public discussion and expression. (p. 120)

Such a reconfiguration means that all social, economic, and political relations *and their interrelations* can and should be theorized. It also makes clear that our definition of political action must be broadened, as well, to accommodate this more heterogeneous conception of public. Abrahams (1992) proposes that we consider "political action a *type* of human behavior rather than as an *arena* (in essence, the formal governmental system) through which people act or attempt to effect change" (p. 329). This view of political action would include actions in a variety of sites or arenas and would include nonelite actions and social relations. Such a redefinition forces a rethinking of political subjects or agents and of coalitional orientations.

Reshaping Political Subjects and Agents

Marxist, neo-Marxist, and rational-choice versions of urban political economy see the subject as unitary—workers, the poor, land owners—rather than as multiple and complex—workers of color and gender, and so on. Theorists recognize this complexity, of course, but for analytical purposes presume economic interests and identities are dominant in shaping choices and actions. But the growing political importance of non-class-based identities and affiliations makes this assumption problematic. Not only are non-class-based identities increasingly important, there is a good deal of evidence that noneconomic values are at least as important as self-interest in influencing everyday actions (Mansbridge, 1990). Yet in most theoretical perspectives, these noneconomic motives are labeled unimportant, irrational, or false.

The way in which self-interest has been defined is related to the public/private dualism as commonly formulated. Love, self-sacrifice, and reciprocity have been assumed to be motives that operate in families and in the home, whereas self-interest is a public motive operating in the market and politics. The abstraction of self-interest in this way reflects a power relation because it gives legitimacy to particular forms of discourse. It fails, however, to acknowledge the social construction of self-interest, wherein the powerful define the terms of debate.

From the logic of feminist and poststructuralist theories, the subject is multiple and contradictory (Harding, 1991, p. 285). If we assume a multiple, complex subject with varied identities, interests, and values, we must recognize knowledge as situated in particular social, spatial, and economic relations and identities. This argues against conventional notions of objectivity and imputed rationality in favor of epistemologies that emphasize differences rather than assume universal tendencies (Harding, 1991). The logical consequence is to acknowledge the potential for positional politics as well. This politics of identity based on differences is beyond the scope of political economy categories, because the multiple identities and transformational features characteristic of movement politics are not easily accommodated by elite-centered approaches until and unless such groups form coalitions that engage with the state (Clarke, Staeheli, & Brunell, in press).

Rethinking Coalitional Orientations

To some extent, urban political economy approaches are limited by their focus on "big issues" of regimes, growth machines, and other versions of elite behavior (Kofman & Peake, 1990). The elite coalitional orientation contributes to weaknesses in dealing with diversity and differences between social groups, both in terms of identity formation processes and social movement activities. On the face of it, this charge seems counterfactual: Most theories emphasize the differential incentives for cooperation and competition across groups and distinguish groups in terms of unique histories, resources, and logics. Yet the underlying theoretical expectation is that groups agree to work together for mutual gain by overriding these differences; the analytic focus is on how groups come together to pursue goals or simply to generate "enough cooperation" to carry out tasks. These coalitional strategies often lead to familiar tensions between coalition leaders and their particular constituencies that tend to create relations characterized by patronage rather than a representation of group interests from the grassroots. The normative and empirical concern of such strategies is to subsume particularistic interests and identities for common gains.

In complex societies, such coalitions are essential, but the form they take is open to question. In an argument akin to Fraser's (1990) notion of strong publics, Young (1990) claims that coalitions based on strong social group representation and autonomy are critical in heterogeneous publics; these coalitions would preserve the identities and interpretations of constituent groups, while still providing mechanisms for structured group representation in decision making. Such coalitional strategies have been identified among some women's political organizations (Boles, 1992). It is notable, however, that some of these coalitions eschew state funding and other interactions with the state apparatus, fearing that such interactions will inevitably force the coalition into subsuming the interests and identities of constituent groups and to more clientalistic relations. As such, these political coalitions carve out arenas for action apart from the state in which they attempt to force social change. By focusing on state apparatus and confounding the state and the public sphere, urban political economy overlooks the nonstate institutional structures

(Wolch, 1990) and self-managed organizations where such alternative coalitional strategies are emerging.

■ Reconstituting Urban Political Economy

The argument to this point is that epistemological and analytic limitations of political economy approaches preclude a full understanding of local politics. In particular, the dualistic definition of public and private, and the consequent elite orientation and focus on organized groups engaged with the state, overlook social differentiation. Yet simply adding "women" or "race" or other social and cultural relations to inherently economistic urban political economy approaches is not sufficient to redress these limitations. We suggest that placing citizenship issues rather than elite arrangements at the center of the debate about local social and economic orders is essential.

The citizenship concept offers a means of bringing together structure/ agency issues to overcome economistic tendencies in urban political economy that often imply a greater role for structure over agency (Swanstrom, 1993). The citizenship concept also shifts the normative and analytic focus to the changing relationship of the individual and the state in local communities. This reconstituted approach balances regime analyses with greater emphasis on citizenship relations as arrangements that mediate economic and political restructuring at the local level. Further, we anticipate that the differential impacts of political and economic restructuring alter the conditions for local political action; to the extent that local political action is critical for constituting local polities, these changes reconstruct the setting for citizenship at the local level (S. Smith, 1989). Thus citizenship is understood to be constructed socially and spatially.

Reintroducing a citizenship focus to urban analyses revives a tradition of political analysis linking the city and citizenship that dates to Aristotle's views. It also places urban analyses in the midst of a resurgent scholarly interest in citizenship, albeit one characterized by considerable debate on the viability of the concept itself. Viewing restructuring processes from a citizenship perspective, therefore, promises new insights on contemporary issues as well as potential contributions to a larger scholarly debate. It is to this notion of a restructured citizenship that we now turn.

Changing Concepts of Citizenship

The scholarly debate on citizenship is extensive and well-articulated. From one perspective, the notion of citizenship is increasingly archaic; to the extent that citizen apathy, disconnection, and alienation indicates a legitimation crisis in the state, citizenship is hardly a meaningful political category much less a centerpiece for political action (Kaufman, 1990). In this view, postmodern politics are increasingly decentralized, eclectic, non-state-focused, and centered on personal identities. Standpoint politics based on personal identities and actions outside the arena of the state are seen as the more appropriate focus of analysis and action and theorization. For this reason, Marxist-feminist scholars have little to say about citizenship.

In contrast, some theorists assume civic identity as prior to all other identities. In the traditional liberal view of citizenship, context matters little: Individuals' needs and capacities are viewed as "independent of any immediate social or political condition" (Dietz, 1987, p. 2). By viewing individuals as bearing rights to equal access to political power, liberal theory divorces citizens from the economic and social conditions that might influence the nature and actual exercise of citizenship for individuals in differing circumstances. This ahistorical, contractual notion of citizenship is challenged on two fronts: by communitarian theorists (e.g., Gould, 1988) who stress the active, collective nature of citizenship, and by scholars who draw attention to the broader social and cultural dimensions of citizenship in both normative (King & Waldron, 1988; Marston, 1990; Shklar, 1991; S. Smith, 1989) and empirical terms (Conover, Crewe, & Searing, 1991; Nelson 1984; Verba, 1990).

The relational arguments emphasize the ways in which social, economic, and political conditions affect citizenship status by altering the public standing of individuals and social groups. Many of these changing conditions are associated with restructuring trends, but urban political economists rarely link these differential impacts to changes in local citizenship status. Similarly, philosophical and theoretical works on citizenship fail to take into account theories of economic and political local restructuring trends. Thus the potential for using citizenship as a concept through which we can analyze the ways in which individuals mediate the effects of political and economic transformations remains unexamined. Constructing urban political economy approaches that are sensitive to citizenship requires more explicit articulation of the relationships between citizenship and place and between citizenship and gender.

Citizenship and Place

Linking citizenship to place is a venerable and continuing tradition of inquiry. Analyses by political theorists assume that scale affects the quality of democracy; political association at the local level is presumed to be more generative of citizenship and meaningful participation than is activity at a larger scale, such as the nation-state (e.g., Gould, 1988; Turner, 1990; Young, 1990). Most empirical analyses refer to ecological features of the locality that affect participation. Differences in size, in political fragmentation, in population mobility, and in labor force participation are assumed to affect the territorial commitment or civic memory necessary for meaningful local democracy (Dagger, 1981; Long, 1991). In analyzing how citizens themselves characterize their status as political agents, Conover et al. (1991) find that views of participation vary cross-culturally and by size of place. Simply put, place matters. Both British and American citizens express more communitarian views of local roles than they do for the nation-state. But it is not clear how or why place matters. Ecological and behavioral perspectives identify conditions for variations in democracy across time and space but cannot explain why these variations exist (Silverberg, 1992).

A contextualized analysis of the relationship between citizenship, democracy, and place should address the relationships and processes that link economics and politics as experienced by different local groups. In this way, attention is directed both to the ways in which identity as a political agent is shaped by changing structural conditions and to the ways in which this identity may be generative of political action. Such an analysis, then, makes it possible to explore the ways in which structural processes are experienced in concrete settings.

The importance of contextualized analyses of citizenship takes on additional meaning under the current rounds of economic restructuring. At a superficial level, it might seem that significance of locality declines in the face of hypermobile capital and new international divisions of labor. But researchers such as Cox (1992) and Preteceille (1990) argue that many analyses of economic globalization provide overgeneralized and insufficient accounts of the effects of globalization on local power relations. They argue that globalism may well generate a new localism:

As the significance of national boundaries blurs, localities may become more important as the settings for mediating the interaction of economic and political processes.

The new localism has two important implications for our understanding of citizenship. First, local conditions assume greater importance in shaping the political, social, and economic context of citizenship. The unevenness of economic and political opportunity structures at the local level poses the possibility of uneven development of citizenship between localities. As a result, a contextualized view of citizenship, in which it is recognized that the concrete expressions of the relationships defining citizenship are varied, becomes essential. The rights, entitlements, and opportunities afforded through citizenship are inextricably tied to place. Second, the new localism is characterized by new arenas of political activism that are driven by changing relations between groups and between individuals, groups, and the state. Identity formation processes are shaped by household strategies for coping with these local changes; because these identity formation processes are prior to group mobilization and participation, local politics cannot be understood without addressing the responses of individuals and groups to place-based experiences of restructuring and changes in citizenship status.

Citizenship and Gender

Bringing citizenship concerns into urban political economy also means recognizing the tension between the formal and informal aspects of citizenship, particularly the political and economic structures through which citizenship is actually exercised. As Pateman (1989), Young (1990), Flax (1992), and others have noted, the claims of equality and an essential "human sameness" constituting liberal political theory promote an inclusionary, formal, political citizenship but overlook the social and economic barriers to full inclusion in the polity. As formal, political American citizenship expanded to include women and slaves (Shklar, 1991), the nature of the "sameness" or political equality achieved by admission to the public sphere was defined by the White, male, Anglo citizens extending the franchise. This gendered concept of equality—same in regard to whom? to what? (Flax, 1992, p. 194)—has forced a concept of citizenship that must either ignore differences or argue that they are not relevant to politics and the public sphere. In doing so, it accepts a notion of politics and citizenship that marginalizes and disadvantages women by disregarding gender differences in the very attributes purportedly constituting full citizenship. For many feminists, the liberal concept of the independent citizen is implicitly gendered; women, regardless of race and class, rarely enjoy the physical and economic independence available to men and thus are excluded from full citizenship.

Because many of the attributes of citizenship in liberal political theory, including independence as a condition for political participation, correspond to conventional dichotomies of the public and private, constructing a feminist view of local citizenship begins with rethinking these dualistic notions. A more inclusive notion of local citizenship must acknowledge the constraints on full citizenship imposed by women's lack of social and economic independence. In this regard, it is essential to conceptualize citizenship as more than the right to vote; citizenship implies the ability to participate fully in all arenas of the polity. It thus becomes important to incorporate a broader understanding of the meaning of "political action." Distinctions between citizen- and community-based politics reflect oppositional dichotomies between the public and private and between state and household institutions that devalue non-state-oriented activities; an inclusive definition of political action would include the nominally private, personal, and informal political processes characteristic of politics at the scale of the household and the neighborhood (Abrahams, 1992). Contextualized analyses are critical to this reconceptualization: Gender relations and gender divisions of labor are intrinsic to economic restructuring processes, as the intersection of global economic and political restructuring is increasingly mediated by local structures of power and inequality.

Gender, Place, and Politics: A Feminist View of Local Citizenship

Our argument views gender relations as critical elements in theorizing about local social and economic changes and sees local citizenship as constituted through relations of gender, place, and politics. Placing citizenship structures at the center of analysis allows for consideration of changing relations of the individual and the state and issues of social difference and identity that increasingly mark local politics and community dynamics. A citizenship focus reestablishes the normative and empirical link between urban theory and democratic thought. For urban political economists, this requires a more contextualized analysis of how gender and place shape problematic citizenship roles.

To account for the multiple, intersecting sites of power contested by individuals and groups, urban political economy must incorporate a concept of citizenship that focuses on the political and economic relations affecting citizenship standing and the ways in which changes in those local relations shape the problematic local citizenship roles of marginalized social groups. A feminist concept of local citizenship is not passive: It acknowledges restructuring processes that change the conditions for citizenship for different groups but recognizes struggles by those marginalized in various sites of power to reconstruct their citizenship.

■ Restructuring Local Citizenship

Given the argument that citizenship incorporates both economic and political relations, we consider three features of local settings that influence changes in women's social, economic, and political standing and thus their local citizenship: (a) changing work regimes, (b) declining social rights, and (c) shifting political access. It will become clear that the effects of each change are ambiguous—offering the possibility of some enhancements and some retrenchments of women's standing as citizens. We argue that the roles of place and other social identities (e.g., race and class) are critical in shaping the outcomes of restructuring for particular groups of women. It will also become clear that these changes are interconnected—that changes in one realm have implications for other realms. Understanding the effects of restructuring on the urban political economy depends critically on the multiple positions of women—and indeed of all residents—within the web of power relations that structure our society.

Changing Work Regimes

The centrality of earning power to the privilege of citizenship can hardly be overstated. As Shklar (1991, p. 68) points out, in American civil society, "one is what one does." To the extent that earning power and employment opportunities are circumscribed, individuals are considered less than full citizens.

Increasingly, work regimes characterized by standardized hours, contracts, wage structures, and work conditions are giving way to nonstandardized work regimes (Christopherson, 1989; Tilly, 1991). In nonstandardized work regimes, the number of hours worked weekly, the work schedule, the location of work, and the contractual relations between employer and employee vary considerably. These changing employment conditions make a decent standard of living a more difficult prospect for many groups. With less discretionary time available, political participation is limited to low cost, minimal strategies, such as phone calls and letter writing, rather than sustained organizational involvement. Further, to the extent that employment conditions reduce perceptions of interdependence among citizens, they are less likely to mobilize and act together on the basis of identity or interest. Thus flexible politics arise from flexible employment: High-wage workers in nonstandardized work regimes have expanded opportunities for political participation, whereas low-wage workers involuntarily in such work regimes have diminished and altered opportunities for political involvement (Clarke & Staeheli, 1993).

Women are disproportionately (although not exclusively) employed in low-paying, flexible jobs (Christopherson, 1989). Recalling Shklar's (1991) argument, women's standing as productive citizens is imperiled by their weak earning power. To the extent that political rhetoric links claims to social rights to productive capacity in the marketplace, women in low-wage flexible employment are vulnerable to further erosion of their citizenship standing. A wealth of evidence shows that women's poverty and economic dependency limits their conventional political participation (Nelson, 1984). But changing work regimes may also create the need for compensatory survival strategies. Some of these survival strategies entail idiosyncratic responses to individual job and household needs; others may involve workers' and households' banding together to provide alternative services on a collective basis. Ironically, some of the very processes leading to the spatial segregation of lowincome households into certain neighborhoods of the city may facilitate the private provision of services and the ability to act collectively. It is not clear, however, to what degree women can use their organization at the community level to press their claims in the larger polity. Thus we have the possibility that citizenship is unevenly developed between localities (as argued earlier) but also that it is unevenly developed *within* them.

Although these initiatives suggest that some groups are overcoming barriers to collective action, the actions of these women may not resemble conventional political activities because they focus on needs ascribed to the private sphere, are grounded in noneconomic bases for action (Mansbridge, 1990), are outside the electoral arena, and often avoid state-funding ties and controls (Clarke et al., in press). Thus changing work regimes can diminish women's citizenship status by limiting their earning power and their opportunities for conventional political participation. But they may also create situations in which women are empowered and enabled to reconstruct their citizenship through collective actions in the community that respond to job and household needs relatively independent of a state that is often seen as irrelevant and unresponsive.

Eroding Social Rights

Marshall's (1950) evolutionary model of citizenship emphasizes the importance of social rights—the ability to earn or gain a decent standard of living—in modern citizenship. Social rights concepts of citizenship also legitimate modern notions of the welfare state. The expanding welfare state brought a more activist government role and greater government penetration into civil society, including a greater emphasis on equality rights in employment and family relations. Overall, welfare policies often involved some redistribution among groups and thus the establishment of ties among citizens who are otherwise remote from each other (Steenbergen, 1992). At the local level, these trends meant an expanded public employment sector (with greater employment opportunities for women), a range of social services provided to households through public agencies, and some fiscal redistribution among taxpayers and beneficiaries. The extent to which women benefit from these social rights is in large measure shaped by where they live, because both the level of benefits (through work in the public sector and through public services and entitlements) and the conditions on those benefits are matters of state and local policy.

Conservative ideology undermines the policy implications of the social rights argument and supports the notion of citizen obligation as a basis for policy choices (Steenbergen, 1992, p. 5). The emphasis is on obligation and productive contribution rather than on ties between citizens or between citizen and government, particularly a public role in providing for citizen's social well-being. As the welfare state is cut back, equality rights, redistributive initiatives, and policy-fostered citizen solidarity ties are jeopardized. In addition, with the shift of many significant decision processes to off-budget agencies and organizations, citizen access to decision arenas is undermined. This depoliticization dampens local political activism and the perceived salience of local politics (Conover et al., 1991). Given these challenges to social rights concepts of citizenship, it is necessary to reconsider the Marshallian notion of citizenship as an individual's membership, or participation, in community and the links between notions of political and social rights.

Recent trends in the economy provide an additional challenge to the welfare state and the concept of citizenship based on social rights. These threats are especially severe at the local level where many services are provided and are subjected to local political and economic conditions. The ability of women to respond to these eroding social rights depends critically on the accessibility of localized political institutions.

Shifting Local Political Access

Even groups able to overcome barriers to collective action may find access to state power limited by their own organizational capacities and by shifting local channels for political access. Groups organized around identity rather than interests often exhibit distinctive organizational structures and decision rules. Many feminists see recent feminist political organization as a model for revitalizing democratic practice (Dietz, 1987). But groups organized around feminist values of nonhierarchical structure, expertise based on experience, and so on, may find political incorporation is difficult when political institutions are organized in hierarchical terms and by decision rules based on majorities rather than on consensus (Clarke et al., in press). Greater incorporation of women and women's voices in governing regimes might lead to a distribution of material goods that benefit women (day care, maternal leave, etc.) but not necessarily to changes in the institutional rules and procedures most directly shaping women's choices and access to power (Young, 1990). The institutional framework of the policy-making process in itself constrains women's effectiveness: One state legislative study indicates that women must work harder to get legislation passed and that their legislation is more subject to amendment, reassignment, and veto than legislation sponsored by male legislators (Kathlene, 1989). Greater inclusion of women elites or representatives of women's groups in a public sphere dominated by men and governed through institutions that are encoded with the values and norms of the dominant group does not necessarily lead to substantive policy change. Even if shifts in the political opportunity structure of the state contribute to group effectiveness independent of internal resource mobilization (Costain, 1992), these disjunctures in organizational structure and processes may be sufficient to limit the voice of some women's organizations.

The voices of women may be further dampened by political restructuring processes. Trends of deregulation, shifts toward off-budget agencies, and provision of services by private and nonprofit organizations signal shifts in access to policy-making processes that may be especially disadvantageous for women. Deregulation means reduced services and benefits for women citizens, who are generally the major beneficiaries of social services and, often, loss of employment by women employed in public sector positions. Although the increased role of private and nonprofit service providers may provide opportunities for more participation and greater accountability (Wolch, 1990), it is accompanied by shifts of "big" decisions on capital and labor to off-budget agencies (Clarke & Gaile, 1989). Both trends limit the accountability of the public sector and the viability of electoral channels; by increasing the costs of participation and access, they skew access to those with the time, expertise, and skills to penetrate these unfamiliar arenas. In an era of flexible politics, local political access is more available to those in voluntary, high-wage, flexible-employment arrangements.

■ Conclusion

In this chapter, we have argued that gender relations are pervasive. Gender relations are implicated in a variety of political, economic, and social relations, even as they shape our epistemological propositions and analytical strategies. The failure to incorporate gender relations into theories has led to partial analyses of urban development. By contrast, a gendered conception of urban politics will enhance the analytic and theoretical potential of urban political economy approaches. Feminist theories offer alternative conceptualizations of the public and private spheres that encourage analysis of the multiple sites of power structuring community life. We demonstrate the potential of this engendered view of urban politics by applying it to an analysis of citizenship; using a relational notion of citizenship, we discuss changes in three sets of relations—or sites of power—that shape local citizenship for women. In our perspective, citizenship standing is mediated by the ways in which gender, place, and politics intersect; although we restrict our discussion to gender relations, the citizenship perspective is germane to analysis of other noneconomic relations, such as race and sexual orientation, not easily addressed through current urban political economy approaches.

The expanded conceptualization of urban political economy can lead to richer theories and understandings of the complexity of urban life. By shifting attention from elites and coalitional strategies in the arena of the state, we can incorporate the agents, motives, and sites that are generative of new political movements as households adjust to changing economic and political conditions. Attention to these issues will enhance the salience of urban political economy to scholars and activists working for social change. In this way, urban political economy can make an even greater contribution to both citizens and our theories of them.

REFERENCES

Abrahams, N. (1992). Towards reconceptualizing political action. *Sociological Inquiry*, *62*, 327-347.

Boles, J. (1992, September). *Local feminist coalitional strategies: Three models*. Paper presented at the annual meeting of the American Political Science Association, Chicago.

Christopherson, S. (1989). Flexibility in the U.S. service economy and the emerging spatial division of labor. *Transactions of the Institute of British Geographers, N.S. 14*, 131-143.

Clarke, S., & Gaile, G. (1989). Moving toward more entrepreneurial economic development policies: Barriers and opportunities. *Policy Studies Journal, 17*, 574-598.

Clarke, S., & Staeheli, L. (1993, April). *The new politics of work and time*. Paper presented at the annual meeting of the Association of American Geographers, Atlanta, GA.

Clarke, S., Staeheli, L., & Brunell, L. (in press). Women redefining local politics. In D. Judge, G. Stoker, & H. Wolman (Eds.), *Theories of urban politics*. London: Sage.

Collins, P. (1986). Learning from the outsider within: The sociological significance of Black feminist thought. *Social Problems, 33*, S14-S32.

Conover, P. J., Crewe, I. M., & Searing, D. (1991). The nature of citizenship in the United States and Great Britain. Empirical comments on theoretical themes. *Journal of Politics, 53*, 800-832.

Costain, A. (1992). *Inviting women's rebellion*. Baltimore, MD: Johns Hopkins University Press.

Cox, K. (1992). The politics of globalization: A sceptic's view. *Political Geography, 11*, 427-429.

Dagger, R. (1981). Metropolis, memory, and citizenship. *American Journal of Political Science, 25*, 715-737.

Dietz, M. (1987). Context is all: Feminism and theories of citizenship. In J. Conway, S. Bourque, & J. Scott (Eds.), *Learning about women: Gender, politics, and power* (pp. 1-24). Ann Arbor: University of Michigan Press.

Flax, J. (1992). Beyond equality: Gender, justice, and difference. In G. Bock & S. James (Eds.), *Beyond equality and difference* (pp. 193-210). London: Routledge.

Fraser, N. (1990). Rethinking the public sphere: A contribution to the critique of actually existing democracy. *Social Text, 25/26*, 56-80.

Gibson, K., & Graham, J. (1992). Rethinking class in industrial geography: Creating a space for alternative politics of class. *Economic Geography, 68*(2), 109-127.

Gould, C. (1988). *Rethinking democracy: Freedom and social cooperation in politics, economy, and society.* Cambridge, UK: Cambridge University Press.

Habermas, J. (1989). *The structural transformation of the public sphere: An inquiry into a category of bourgeois society* (T. Burger, Trans.). Cambridge: MIT Press.

Harding, S. (1991). *Whose science? Whose knowledge? Thinking from women's lives.* Ithaca, NY: Cornell University Press.

Kathlene, L. (1989). Uncovering the political impacts of gender: An exploratory study. *Western Political Quarterly, 42*, 397-421.

Kaufman, L. A. (1990). Democracy in a postmodern world? *Social Policy, 21*, 6-11.

King, D. S., & Waldron, J. (1988). Citizenship, social citizenship, and the defence of welfare provision. *British Journal of Political Science, 18*, 415-443.

Kofman, E., & Peake, L. (1990). Into the 1990s: A gendered agenda for political geography. *Political Geography Quarterly, 9*, 313-336.

Lawson, V. (1992). Industrial subcontracting and the work-welfare relationship in Latin America: A framework for contextual analysis. *Progress in Human Geography, 26*, 1-23.

Long, N. E. (1991). The paradox of a community of transients. *Urban Affairs Quarterly, 27*, 3-12.

Mansbridge, J. (1990). *Beyond self-interest.* Chicago: University of Chicago Press.

Marshall, T. (1950). *Citizenship and social class.* Cambridge, UK: Cambridge University Press.

Marston, S. (1990). Who are the "people"?: Gender, citizenship and the making of the American nation. *Environment and Planning D: Society and Space, 8*, 449-458.

Nelson, B. J. (1984). Women's poverty and women's citizenship: Some political consequences of economic marginality. *Signs, 10*, 209-231.

Pateman, C. (1989). *The disorder of women: Democracy, feminism, and political theory.* Stanford, CA: Stanford University Press.

Preteceille, E. (1990). Political paradoxes of urban restructuring: Globalization of the economy and localization of politics? In J. R. Logan & T. Swanstrom (Eds.), *Beyond the city limits* (pp. 27-59). Philadelphia: Temple University Press.

Shklar, J. (1991). *American citizenship: The quest for inclusion.* Cambridge, MA: Harvard University Press.

Silverberg, H. (1992). Gender studies and political science: The history of the behavioralist compromise. In J. Farr & R. Seidelman (Eds.), *Discipline and history: Political science in the United States* (pp. 363-381). Ann Arbor: University of Michigan Press.

Smith, D. (1987). *The everyday world as problematic.* Boston: Northeastern University Press.

Smith, S. (1989). Society, space, and citizenship: A human geography for the "new times"? *Transactions of the Institute of British Geographers, 14*, 144-156.

Steenbergen, M. (1992, September). *The citizenship basis of public welfare attitudes: Issues of theory and measurements.* Paper presented at the annual meeting of the American Political Science Association, Chicago.

Swanstrom, T. (1993). Beyond economism: Urban political economy and the post-modern challenge. *Journal of Urban Affairs, 15*(1), 55-78.

Tilly, C. (1991). Reasons for the continuing growth of part-time employment. *Monthly Labor Review, 114*, 10-18.

Turner, B. S. (1990). Outline of a theory of citizenship. *Sociology, 24*, 189-217.

Vaiou, D. (1992). Gender divisions in urban space: Beyond the rigidity of dualist classifications. *Antipode, 24*, 247-262.

Verba, S. (1990). Women in American politics. In L. A. Tilly & P. Gurin (Eds.), *Women, politics, and change* (pp. 555-572). New York: Russell Sage.

Wolch, J. (1990). *The shadow state.* New York: Russell Sage.

Young, I. M. (1990). *Justice and the politics of difference.* Princeton, NJ: Princeton University Press.

2 Defining Feminist Community: Place, Choice, and the Urban Politics of Difference

JUDITH A. GARBER

■ The "Community Problem"

Community poses a problem for feminism because too many political theories romanticize traditional community forms oriented toward place. Feminist critiques of this "community problem" have exposed the gender, class, and racial implications of various community concepts. As a result of this inquiry, local *communities of place*—the "family, neighborhood, school, and church web" (Friedman, 1989, passim) prescribed in past and current communitarian theories—are counterposed against *communities of choice*—friendships and identity groups consistent with diversity and feminist ideals (Friedman, 1989, pp. 286-287; see also Young, 1990b, p. 172). I argue that this dualistic framework fails to capture the depth or range of women's community relationships, and it misses the feminist potential of democratic places. One possible escape from the limits of this dichotomy are *communities of purpose*, where shared situations foster local political action. Donning feminist analytical lenses requires us to keep a critical distance from traditional community notions, because there are real dangers in a blind devotion to place-based social and political forms. To the extent that community is identified with localities, the feminist critique also impinges on urbanists' fundamental concern with place. However, critiques of community must appreciate the findings in urban research about how women approach the challenge of creating space for democratic, diverse communities in the places they inhabit.

AUTHOR'S NOTE: This chapter has benefited from the comments of Annalise Acorn, Susan Clarke, and Valerie Lehr, and the research assistance of Harold Jansen.

■ The Flight From Community

Feminist distress about communities of place is directed at democratic theory and practice in which an ongoing community with settled, embedded values organizes political life. Most communitarians endorse smaller rather than larger governments and direct citizen participation over representative schemes, to the extent feasible today. Generalizing about the features of communitarianism is risky, however (Fowler, 1991), because the label encompasses all those—on the left and right—who accord priority to the community over individuals and groups (see MacIntyre, 1981; Michelman, 1988; Sunstein, 1988). The most intellectually coherent and familiar strain of communitarian theory is civic republicanism, traceable from classical times, through Rousseau to the American Jeffersonian and Canadian Tory traditions. A crop of recent philosophers continues to voice the essential civic republican sentiment that politics and citizenship are predicated on "individual . . . fulfillment in relationships with others in a society organized through public dialogue" (Bellah, Madsen, Sullivan, Swidler, & Tipton, 1985, p. 218; see also Barber, 1984, p. 153; Sandel, 1982, pp. 173-174; Taylor, 1985, pp. 213-214).

Liberal feminists, like all liberals, worry that the natural rights of *individual* women are subsumed in strong-community schemes (Gutmann, 1985). Another, distinctive danger is that women as a *group* or set of *groups* suffer under close-knit, authoritative, and homogeneous communities of place, whether real or imagined. Too often, such arrangements produce systematic exclusions and oppressions, not the least of which is patriarchy.

This second complaint about community is integral to a new wave of feminist theory sharing a broadly deconstructionist perspective, which undermines the modernist search for universal understandings, objective facts, or essential qualities. Rather, deconstruction of social and political life involves recognizing contingency, subjectivity, and dissonance. Regarding feminism, then, no unqualified concept such as *women* is valid, and a woman's interests can be defined only from her own experience of her social position (Butler & Scott, 1992; Nicholson, 1990). In this literature, there is also a pervasive argument for shifting political imperatives away from their traditional territorial and governmental referents toward group interests articulated from the standpoint of race, sex, and sexuality. Postmodern (Hekman, 1992; Phelan, 1989), "difference politics" (Young, 1989, 1990a, 1990b; see also Friedman

(1989), and poststructural (Mouffe, 1991) critiques of communitarianism proceed along these lines, although they do not espouse identical opinions about community.

This line of argument is trenchant, particularly in its skepticism about the appropriateness of attempting to reach the "common understandings" (Phelan, 1989, p. 156) communitarianism requires. However, key feminist deconstructionist treatments of community suffer from a highly constricted conception of the central subject of their inquiry, taking as given what is actually at issue. Except for theorists who believe females naturally form the "connected" relationships necessary to communal politics (Dietz, 1985; Sherry, 1986), feminists much too readily would give up the ability to define community to the philosophers who have presented disappointing models of it. At the same time, they contest the ideological appropriation of community by White men. Current feminist theory tends to present women a stark choice between suffering bad political and social arrangements and rejecting local community altogether. Valuing such communities is not widely acknowledged to be a viable option.

The flight from defining place-based community is a contradiction, as a prominent theme of deconstructionism, and what most clearly differentiates it from modernist liberal, Marxist, and communitarian theory is that politics cannot proceed when the ineradicability of conflict is denied in favor of majority rule or consensus. Because this is a cautionary note well worth heeding, taking local community as given is particularly unsettling. It appears that as long as women cannot insert themselves into the dominant ideology, community is judged a fundamentally flawed idea rather than one that has worked badly for women and could be rescued if conflicts over community, citizenship, justice, and so on were brought to the fore.

This is a theoretical issue, but critics of communitarianism also do not take care to understand how actual women regard community, overlooking a crucial fact: Women participate daily in "ongoing ideological conflict over the meaning of place" (Hummon, 1990, p. 28) in their struggles to make community their own. In certain respects that I will discuss, local places succeed at providing meaningful, regular opportunities for women who may not otherwise understand their lives as having political dimensions to express their multifaceted desires for relationships in a public sphere. These relationships are alive in the denser central-city settings where women—and above all poor women—increasingly find themselves, compared with men. It is unclear where women would prefer to live if they had real choices (see Shlay, 1986), but feminist urban research conducted over two decades demonstrates that there is a bond between community and city for many women. This implies that feminist political thought is not driven by an expansive enough view of community or a deep understanding of cities.

Social science work also reveals how cities are pervasively gendered, to the disadvantage of women at work, at home, and in public spaces. It moves beyond urban analyses (whether in geography, sociology, or political economy) that treat individuals abstractly with regard to their sex and hence make men the universal standard for urban dwellers. These findings about cities have not, however, been adequately applied to the sexual inequalities embedded in the institutional relationships of governance, citizenship, and community.

Thus we have two bodies of feminist knowledge at hand: political theory and urban analyses. Each is incomplete, but they have obvious sympathies. Together, they set the stage for a distinct, feminist viewpoint on locality and can help determine whether cities (in theory and real life) must be cut loose from their traditional community moorings or if place-based community can be sufficiently redefined to meet feminist goals.

■ Community as Fraternity

Because it is a theory of public virtue, not of private interests, civic republicanism purports to be more democratic and egalitarian than liberalism. In reality, strictly communitarian local institutions are no more likely than liberal ones to respond democratically to women (and people of color and poor people). Radical transformation of gendered politics is difficult because, as Pateman shows, both models of citizenship and community are patriarchal: For liberals, public freedom for individual men rests on the private (sexual) subordination of women as a class (Pateman, 1988, 1989, pp. 210-220), whereas communitarian politics delves into the private sphere to ensure more equal political participation only insofar as men's inequalities are concerned (1989, pp. 220-222; see also Mansbridge, 1980, p. 94). In neither case do political mechanisms provide for women's consent to the sexual division of power according to which chances for meaningful citizenship are allocated.

Like liberals, communitarians treat the city instrumentally, although they vaunt the communal life and may include cities in that scheme (Barber, 1984, pp. 261-311; but see Taylor, 1985, p. 229, n. 37). Each theory treats cities as means to valued ends that are abstract as well as predetermined. For liberals, the end is the self-determination of individuals, whose interests are formed outside of politics; for communitarians, the end is self-government of the civic whole, whose good is formed via participatory politics. Each resists awarding primacy to social diversity, physical and material security, healthy environments, and "androgenous" (Saegert, 1988) spatial design, in whose absence neither feminist ideals nor viable cities can flourish (see Andrew, 1992). At the intersection of feminism and cities, then, it is necessary that such conditions be treated as vital, concrete prerequisites of democracy—not abstractions—and that they be carried out via processes sensitive to political and social contexts—not predetermined rules.

Community is steeped in fraternal imagery and practices (Hekman, 1992, p. 1109; Young, 1989, p. 253) that are most damaging because they conjure up something that has always been largely mythical. Pacific small towns and manicured suburbs can exact a high price from women and other outsider groups. On a crude level, because interlopers may breach the consensual nature of public decision making that is predicated on the sameness of the citizenry, they can justifiably be shunned or silenced. Strangers might so threaten the community's perceived identity and economic self-interest (Rayside, 1989, pp. 113-118) that certain people are denied entrance or driven out, often legally. Besides techniques to exclude subsidized and rental housing, people of color, group homes, or untraditional families, violence by neighbors and police is not unheard of.

More subtly, a set democratic process and preexisting norms aimed at fashioning political agreement are taken to be necessary for perpetuating the community (Barber, 1984, pp. 135-136). They are not themselves usually open for discussion or contestation, and it is not coincidental that detailed procedural outlines dominate current republican theorizing. Indeed, attention to procedure is entailed by the search in democratic communities for principles such as "universalism" and "political truth" (Sunstein, 1988, pp. 1554-1555; but see Barber, 1984, pp. 129-131). As in more intentional forms of exclusion, the requirements of the whole conspire against honoring basic objections to the community will, such as stating that there is unequal treatment of women by men and of women by other women or (distinctly more subversively) that genuine consensus is impossible or fictitious. As I mentioned earlier, liberals have always insisted that the harmonious public sphere oppresses individuals. It more readily silences people as members of groups—as women, young or elderly people, gays and lesbians, people of color, or immigrants (Bell & Bansal,

1988). This recalls the parochialism and intolerance most frequently associated with local autonomy in the United States but that also exist in Canada (Rayside, 1989) and Europe. American communitarian legal scholars take pains to rescue the tradition from itself by preserving pluralism in the democratic process (Michelman, 1988, p. 1495; Sunstein, 1988, pp. 1574-1581), but in doing so they end up tempering civic republican tenets with a strong dose of liberalism (Sullivan, 1988).

In recent political discourses, patriarchal and xenophobic structures are invoked overtly to rationalize the dominant configurations of power and opinion. As a strategy in itself and also a result of that strategy, local community is conflated with "traditional" values inimical to the majority of women and all feminists (Friedman, 1989, p. 281). Part backlash against efforts to ratify and encourage pluralism alongside the political and cultural mainstream and part reassertion of the inherent conservativism of sanctioning "an image of family-like communal cohesiveness" (Rayside, 1989, p. 118), the explicit linkage between the dominant societal ideology and community has merely accelerated feminist fears.

Finally, there has historically been a gendered division of political labor in local communities that gives the lie to certain classical justifications for communitarianism. Female local activism, although true political leadership in every sense, has tended to mirror women's domestic concerns—housing, child care, welfare, safety, the environment. In a sense, these activities amount to "public housekeeping" (Morgen, 1988, pp. 111-113; Saegert, 1988, p. 29), in which women organize to take up the slack left by "city fathers" (Bellah et al., 1985, p. 170) but with fewer of the political and economic resources conferred by the state, particularly for women who are not White and middle-class. To the extent that women furnish the majority of (unpaid) labor as the conscience of the community, indulging state—that is, men's—animus or inattentiveness toward ensuring what most feminists would regard as an acceptable quality of urban life, it is doubtful whether communal democracy lives up to its advertised ability to bring the public and private spheres closer, create an engaged and empathetic citizenry, and promote social obligations (Mansbridge, 1980, pp. 97-114, 183-209).

■ Difference and the Myth of Common Understandings

Community need not be fraternal to be oppressive or exclusive, because women differ by race, class, immigration status, language, religion, sexuality, marital status, age, and physical ability. The desolate material conditions of so many urban women's lives—which worsen as the late 20th-century economic, geographic, and social shifts exacerbate the division of women into winners and losers—concretely manifest these differences. Because women are not positioned equivalently with respect to structures of power and privilege, proclamations about the existence of a unified *feminist* civic sphere are suspect. In this context, the benefits of community cannot redound to all women, and what constitutes community or benefits will itself be deeply controversial.

Local communities that satisfy the primary conditions for feminist politics—diversity and equality—are therefore elusive. This realization evokes an uncomfortable question: Will feminist efforts to create political communities of place invariably founder on the shoals of a misplaced or even pernicious desire for "shared final ends" (Young, 1990b, p. 238) in the civic sphere? One answer is provided by deconstructionism, which helps debunk the myth that anything in politics can be genuinely common across an entire jurisdiction. As Phelan (1989) points out, "The postmodern suspicion of our intersubjective world deprives us of the easy recourse to common understandings" (p. 156).

But as we will now see, feminist deconstruction of cities and communities comes at the price of depriving us of the recourse to political understandings—admittedly grindingly difficult—built on conflict but that are true understandings nonetheless. Positing "difference" as the primary descriptive and normative window on local political life sets communities of choice or identity groups against communities of place. This rigid framework tends to "remov[e] the grounds for political action" (Phelan, 1989, p. 151) by "removing the ground for negotiations about judgments" (p. 158).

■ Communities and Cities

On the street and in the academy, city is interchangeable with community. This convention makes good sense, not because the two are coterminous, but because they are overlapping and interdependent. Perhaps the most important effort within feminist theory to disengage them has been articulated by Iris Marion Young (1990b; see also Friedman, 1989; Young 1989, 1990a). Young (1990b) argues that justice is hindered by community and locality, but not necessarily by cities, conceived properly. "As an alternative to the ideal of community" proffered by civic republicanism she offers "an ideal of city life as a vision of social relations affirming group difference" (p. 227). The vast potential of cities to foster nonexclusive social differentiation, variety, eroticism, and publicity (pp. 13, 238-240; see Friedman 1989, pp. 286-290) is what constitutes "city-ness" for her. It remains only "an unrealized social ideal" (Young, 1990b, p. 227), because cities have cultivated the worst characteristics of both communitarianism and liberal capitalism.

Size

As a general rule, the larger the city, the better chance that diverse identities will be honored instead of being merged into communal norms or forced underground (compare Young, 1990a, p. 318). Like Jane Jacobs (1961), whom she draws on, Young believes that metropolises counteract the conformist pressures within smaller communities. In contrast, Young does not believe that city identity—community, in other words—is a positive political value. Urban places, defined as spaces in which anything and anyone goes, would be ideal if the complicating factor of local governments and communities were eliminated. In their absence, the politics of difference could flourish; this major amendment to civic-oriented political theories is integral to her feminist vision. For these reasons Young (1990b, pp. 252-255) proposes restructuring cities into enormous regional units capable of regulating diversity-maintaining mechanisms that municipalities and neighborhoods lack. Neighborhood assemblies would be represented regionally and "have autonomy over a certain range of decisions and activities . . . only if [they] do not" harm, inhibit, or dictate the conditions of individual agency (pp. 250-251). In neighborhood assemblies, participants would learn by the "discussion and implementation of policy" (p. 253) made at a higher level to validate group differences and true (nonliberal) equality among identity groups (p. 248).

Although some urbanists would concur that autonomous, small jurisdictions obstruct democracy and should be supplanted by decisions from the center, Young's overall picture of city life is flawed. If we take seriously Jacobs's (1961) admonitions to urban planners, feminists might reasonably conclude that smaller units that invariably resolve themselves into communities based on geographic, social, and/or cultural proximity are prerequisite to successful metropolises. Jacobs herself argued that cities must encompass three sizes of "neighborhood" units, each retaining certain powers of self-governance (p. 117). Particularly at the street scale, neighborhoods provide a measure of safety out of proportion to their size (p. 119), which is no mean achievement. Contrast this with Young (1990b, p. 237), for whom the most prominent attribute of the city is the cloak of anonymity it offers those who would stand out as nonmembers of the monolithic community, a place where neighbors who watch out for you would merely be nosy. Identity groups based on attributes such as sex, race, or sexuality could serve to bridge the gap between anonymity (or even benign neglect) and social connectedness that virtually everyone needs for safety and sanity, especially in the big city. But the subtler conceptual distinctions between community and identity groups—or interest groups, for that matter (Mouffe, 1992, p. 380)—threaten to slide into semantics (see Young, 1990a, p. 320), in part because the neighbors hoped for by communitarians and feared by difference theorists are both caricatures.

Separatism

Historical attempts by women to cultivate communities (or communes) friendly to feminists, lesbians, women of color, and/or poor women illustrate further dilemmas facing identity groups in the city. In these examples, community is a territorial unit—people who live together in a city, neighborhood, or housing complex—and *at the same time* an identity or affinity group unrelated to geography or governmental boundaries. Concrete benefits come from group self-determination and separation from a hostile society: Separatism has liberated and saved the lives of women who in larger society are cut off from nurturing groupings of women with shared identities (Anderson, 1994, pp. 437-438; Phelan, 1989, pp. 52-54); bonding and solidarity motivate female-only education and housing. Nevertheless, success depends on surrendering the ideological battle over citizenship and political community, and separatist strategies are at odds with feminist political agendas aimed at collective, coalitional action for wide-scale social change (Phelan, 1989, pp. 168-169) or changing the priorities of the larger community. It is not that separatism is unpolitical, of course, but that its politics of introversion and homogeneity pit the community against the outside world rather than providing a bridge to like-minded others (but see Anderson, 1994, pp. 443-447).

Political Action

Both identity and anonymity (Haeberle, 1993, p. 2; Young, 1990b, p. 238) are valued urban attributes, but they work at political cross-purposes. Hence the political implications for cities of the distinction between an anticommunitarian pluralism and a community that requires pluralism are great. Just as marginalized residents of cities are unlikely to gather their political strength spontaneously, affirming group difference without striving for (figurative) shared ground on at least some issues is the ultimate estrangement of political empowerment and action. It is unsurprising that Young (1990a) defines what remains of locality without political referents, envisioning cities as "vastly populated areas[s] with large-scale industry and places of mass assembly" (p. 317) and regional governments as "units of economic interdependence, the geographical territory in which people both live and work, in which major distribution occurs" (1990b, p. 252). Conjuring this sterile city helps illustrate that, politically, city and community are mutually dependent, and it feeds the suspicion that banishing effective community from the local scene out of deference to (possibly) unreconcilable difference turns cities into branch plants of the central state. Community can make cities sites of political action, whereas decorative neighborhood assemblies underneath structures of "democratized regional planning" (p. 253) are unlikely to radicalize anybody.

■ Place and Choice

Jacobs (1961) presaged certain primal fears about the communitarian ideal currently expressed by feminist critical theorists, but she did not shrink from community. She argued that

> for all the innate extroversion of city neighborhoods, it fails to follow that city people can therefore get along magically without neighborhoods. . . .
> Let us assume (as is often the case) that city neighbors have nothing more fundamental in common with each other than that they share a fragment of geography. Even so, if they fail at managing that fragment decently, the fragment will fail. . . . Neighborhoods in cities need not supply for their people an artificial town or village life, and to aim at this is both silly and destructive. But neighborhoods in cities do need to supply some means for civilized self-government. This is the problem. (p. 117)

Notably, Jacobs uses neighborhood with precisely the same level of generality and substantive implications that people normally connect with the term *community* but without the dichotomy between place and choice that is a pillar of the feminist "difference" critique of communitarianism. At base, this dichotomy is false and misleading.

First, communities are not merely containers holding people who share a unidimensional identity. (Leave aside the elementary point that all women have multiple, conflicting identities that are no more guaranteed easy expression in a group than in a place.) In neighborhoods and localities, identity and place *necessarily* shape one another's character, because territorial communities are little more than the people who make them up, plus society's reaction to those people. The interaction between place and people is one reason why an impoverished African American community in rural Georgia (Greene, 1991) is politically and culturally miles away from an impoverished African American community in urban St. Louis (Boyte, 1984, pp. 95-114), why the initial gay politics of AIDS was radically different in New York and San Francisco (Shilts, 1988), and also why places change.

Case study research demonstrates in rich and varied detail how the intersection between women's collective attention to place and their collective concern for a particular class, race, or ethnicity fuels their political strength (Bookman & Morgen, 1988; see also Martinez-Brawley, 1990, pp. 47-50). Although some women form coalitions across identity lines to secure joint benefits for the local place in which they live, one's sense of identity might well encompass people far beyond the neighborhood or city limits. Especially for those with few privileges, nationalist identity is a counterweight to the dominant culture. Grassroots work in the urban arena cultivates political efficacy and identity consciousness, and it forges links to national and international social movements directly relevant to women and children (Gittell & Shtob, 1980, pp. S73-S77). Such "strateg[ies] of diverse but pragmatic affiliations" (Gilkes, 1988, p. 55) need not divert focus from the local community; quite to the contrary, solidarity on several levels may actually make local action more attractive and significant. Where the fortunes of one's neighbors and friends are seen as being connected—if they are connected—the stakes of political action become very high. Viewed this way, the community problem is that there are too few of them.

Second, people gravitate to particular geographic locales in part for togetherness, so one's community of place and community of choice may coincide (Friedman, 1989, pp. 284, 290). The point can be more easily seen within cities than between them, because intercity movement carries higher costs. Neighborhoods and districts are not formed randomly but are to some extent composed of self-selected residents who have numerous reasons for their choices and who may be choosing a comfortable identity fit or diversity. City dwellers, especially, often experience community primarily via the general type of plurality that Young advocates (see Hummon, 1990, p. 174). That is, they are connected to and willing to contribute to the "good city" because they do not view themselves as members of one group to the exclusion of others. Complex political self-images permit them to conceive of a community that exists beyond either the merger or the centrifugal force of its parts. This is not to deny that great barriers to free selection of places exist, that some moves are involuntary, or that not everyone desires or engages in community. People often flee constricting communities of origin and of happenstance as soon as possible. But these people go to *places.* Therefore, it exacerbates the community problem to, in essence, reduce geographic communities to "dependent children, elderly persons, and [others] whose lives and well-being are at great risk" (Friedman, 1989, p. 290), immobile and lacking alternatives. It "denies people their roots *in* communities" (Ackelsberg, 1988, p. 302), which they might carry with them from their place of origin for cultural and emotional survival.

Third, occupying the same space supplies people with shared experiences, is a de facto commonality, and may create political cooperation among unlike individuals. For instance, although gentrification exerts pressures on housing affordability and neighborhood stability felt by poor and near poor "incumbent" female heads of households, female gentrifiers are economically more "marginal" than men, their household composition is closer to that of female residents (i.e., single women and single mothers) (Rose & Villeneuve, 1988, p. 54), and there is evidence that women simply place a greater value on urban amenities (see Wekerle, 1988b). Therefore, women *may* discover common political ground around place-based issues, such as the quality of neighborhood schools, the safety of streets and parks, and the adequacy of public transportation (Rose & Villeneuve, 1988, pp. 51-57).

Descriptive and Normative Frameworks

In short, a place-based versus an identity-based conceptual framework community fails on descriptive and analytical grounds, just like modernist urban dualities (see Beauregard, 1993). In terms of simple description, the framework is too rigid to describe all the communities women inhabit and seek out. Although finer distinctions between communities—state based, interest based, chosen, discovered, temporary, enduring—are allowed for, they are sorted dichotomously, too, and end up as secondary categories of the primary division, whereas in reality none are mutually exclusive or necessarily incompatible. In terms of analysis, a dualistic approach to community begs the question of how places and interests interact, which is perhaps the central issue in urban politics.

The dichotomous conceptual framework falls short as a normative hierarchy—identity trumps place—because there is no convincing method of ranking community concepts on practical or moral grounds. None guarantees democracy, if real-life communities are an indication, and it is indeed taxing to imagine a strong normative theory of politics or of cities in which the distinction between abstract communities is a key principle. For instance, there is really no reliable method for distinguishing (on defensible normative grounds) an inner-city neighborhood of poor mothers fighting a toxic waste dump from a suburban community of White middle-class mothers fighting the same thing, unless perhaps the choice is between the two. In the latter case, which is common, the harms and benefits of urban life must be more equitably dispersed than at present, which requires that communities of all sizes adhere to basic rules about not poisoning people. But this rule is not inherently related to the value of place-based community, an issue with prominent spatial elements. Space should not be "objectified as a perpetrator of inequality" (Wilder, 1993, p. 407), because it cannot be judged independently of the exclusionary "institutional processes [that] operate in space" (p. 408). Although they surely interact, injustice and compensatory justice (Young, 1990b) are not dimensions of territorial organization.

■ Feminist Community in Practice

Feminists are the last assiduous practitioners of the basic tenets of communitarian politics, an irony that has not escaped feminist theorists (Pateman, 1989, p. 220; Young, 1990a, pp. 301, 312). Despite predictable, painful tendencies toward exclusion and inattention to power differences between women, feminist political communities nonetheless encompass a wider range of collective political practices than communitarianism's rigid insistence on consensus would lead one to expect. The *idea*, not to be confused with any *ideal*, of community is a motivating force in far-flung urban places, kept alive by women for whom it is their only source of political empowerment.

Feminist research on cities refers frequently to community, and this focus is most stunning when female urban activists speak for themselves (e.g. Boyte, 1984; Gilkes, 1988; Luttrell, 1988). This observation alone hardly justifies hitching the feminist political wagon to local places, although it is hard to ignore that they seem relevant and compelling for even those marginalized urban women who are supposed to benefit from the demise of communities of place. Why do women gravitate so easily to this form of community, given the feminist criticisms? The answer is that, as an abstraction, local community is deeply problematic; in practice, it may actually serve women more often than we think. Feminist critiques tend to take the opposite tack, arguing against the theory by illuminating the horribles of real life. The benefits of creating some shared public life explains why, when the formal political community of the city betrays women, they create their own. Feminist political theory has, rightly, pointed out that shared public lives have profound limits. Even so, imperfect community appears to go farther toward meeting some women's needs than nothing at all.

Housing is the intersection of community, locality, and domesticity, and it best exemplifies women's interests in linking place and politics. Women's housing cooperatives in several Canadian cities demonstrate that the prospect of building community from the ground up, incorporating certain feminist values and the requirements of residents, attracts "powerless" women to intensely political endeavors. Wekerle's (1988a) account of the co-ops portrays models of community substantially like cities in their governing responsibilities but different from patriarchal communities in process and ultimate aims. Women joined the co-ops not only for affordable shelter but also to attain certain political ideals, mutual support, and, according to a resident, "to share responsibilities, rights, democracy" (p. 108). Tenants make decisions about allocating space between communal and private uses, budgets, service delivery (including child care and "special needs" residents), membership (the inclusion or exclusion of particular groups), policy making and enforcement, management, redistribution, and economic development, *in an urban setting*. Leavitt and Saegert found similarly rich, multidimensional communities in New York City housing projects among very poor female heads of households, who were overwhelmingly women of color

(Saegert, 1988, pp. 34-35), and in Chicago, single-room occupancy (SRO) hotels have served for a century as diverse urban communities for the near homeless. Bound by kinship, friendship, and neighborhood ties, SRO residents help each other out but retain a fundamental independence (Hoch & Slayton, 1989, pp. 147-171). The community practice of "mutual responsibility" (Hoch & Slayton, 1989, pp. 240-254; see also Sypnowich, 1993, p. 499), rather than merger, may be put to political use to preserve SROs threatened from outside.

In the "community aspects" of these housing arrangements, individual women discovered a relatively safe place for themselves, compared with more extensive public spheres. And this happened while engaging in fundamentally urban tasks. Localities can ensure individual/group and community needs simultaneously if they do not also isolate women by aiming for unity or support the oppression of some women by other women under the guise of community. The trick is figuring out how to balance these competing tendencies within local social groupings. Grand theories help us understand more about the oppression perpetuated by exclusion and political inattention to the differing positions of people who make up communities. A more grounded focus would acknowledge that it is a particularly awkward moment for feminists to dissociate the urban from the community in their efforts to theorize a nonexclusive political framework. In the United States, a search for roads leading back to functional communities of place occupies residents of inner-city neighborhoods and, particularly, the women who are their backbone.

■ Communities of Purpose

Politics makes little sense if it is not conceived as a joint effort—it is inescapably a *community of purposes* and sometimes also of interests. Communitarianism remains antithetical to feminism, however, as long as theorists, even those sensitive to issues of plurality and oppression, continue to think that communities are composed of the "people," who formulate and voice "our" (Michelman, 1988, pp. 1524-1532) consensual values and ideas (Hekman, 1992, p. 1115). Locally, as globally, conflicts over differences degenerate into violence and repudiate "the community" of interpersonally generated understanding.

On the other hand, focusing criticism on "the privileging of face-to-face relations by theorists of community" (Young, 1990a, p. 313) somewhat misses the point of cities. Face-to-face relations are part of urban life, like it or not. They may not be happy relations, but neither are they anonymous. Caught between the rock of claustrophobic local communities and the hard place of the elusiveness of achieving invisibility through group identification, perhaps the best that can be hoped for is a wary truce; perhaps this is the basis for communities of purpose. For the practical purposes of action and self-help, a commonality of places and situations permits differences to be recognized, not ignored *or* (falsely) overcome. Communities of purpose for mutual responsibility and cooperation are integral to feminism and democratic cities if political action is a goal.

Accepting that communities empower some women and materially improve their lives means accepting certain manifestations of the flawed communitarian vision. There is, however, a crucial difference between a hegemonic community and the matrix of communities that serve as the political and cultural anchors for many women (and men) who dwell in urban and suburban cities. In the first case, the republican notion of the public sphere threatens those who do not already have power—it simply cannot create political power where there is a deficit for reasons outside the political process or where the process itself is being challenged. In the second case, a meaningful public sphere encourages alternate political discourses that dissent from the majority and conflict with the dominant community's procedural and substantive agreements. If the communities that "we" (Mouffe, 1991, pp. 79-80) attempt are understood as prerequisite to conditions of equality and diversity, and not as mirrors of some preconceived notion of these terms, then collective action and communities of purposes located in places make good sense.

Feminist critiques of place-based community will invariably fail urban theory if they take civic republicanism on its own terms and give in unwittingly to its homilies about community intimacy and consensus. Rather, rights, remedies, and other instruments of individual self-interest may be the only rationale for political concern for people long denied their protection; participation may be sporadic and strategic. The conditions of local politics presented by communitarians are defective and not definitive, and accepting them is the flight from community presented earlier.

However, feminist theorists of communal life must think as deeply and in as much detail about democratic process and institutional design as do communitarians. Process sets the terms for participation by stating how differences and histories of oppression must be accorded validity in setting community values. No wonder feminists concerned with radical democracy *as well as* deconstruction (Mouffe, 1991; see also Cornell, 1987; Sypnowich, 1993) discuss community compellingly, for democracy demands attention to political action by equals and how this might be achieved. Individuals identify with a community because its own identity and political direction derive from them; its legitimacy is produced collectively, not imposed without regard to the requirements of the citizens. Collective action toward social change is unthinkable outside the context of relationships between people who believe their well-being is tied together, for politics is not a singular endeavor. Contrary to the claim that feminist efforts to reconcile individualist and communitarian theories are fruitless, because both are patriarchal (Hekman, 1992, p. 1113), no patriarchal theory could sustain a conception of community where politics radically transforms self and society.

What these theorists fail to discuss is what is most important in cities, rules about sharing space. Nonetheless, the principles expressed at the juncture of critical feminism and radical democracy apply to local communities of place and to the question of how shared space can be organized and governed in a nonexclusionary way. Radical democrats necessarily conceive of a politics larger than localities, and a broadly "left" movement that brings separate, spatially located politics together—but does not diminish their local importance—is indispensable. No such force exists; therefore, feminists and urbanists must say how local and larger politics might join up. Until feminists recognize that place-based community is pivotal to the politics of shared situations, this type of theorizing is unlikely.

■ Contesting Community

It is true that "American culture incorporates different, contrasting community ideologies" (Hummon, 1990, p. 169) (this applies equally to Canada), but the ideological clash has not been robust enough to institutionalize in local state-sponsored political processes the voices of marginalized groups. Until community discourse acknowledges women's versions that have been excluded from defining the "official" community or relegated to housekeeping detail for city politics, feminists will avoid communitarian theories for advocating political and social organization inhospitable to the differences that define people.

However, abandoning place-based community is not the only possible feminist stance. First, advancing the variety of social and economic issues that affect women entails political alliances extending beyond the group, neighborhood, city, or nation. Broader exercises of citizenship are likely to occur only when organization and empowerment are first practiced on at least one level of community, and the locality has historically been the level most accessible to women. Second, many of the problems that female city dwellers experience *as women*—homelessness, violence against and by children, sexual assault, lack of public transportation, economic ghettoization—are indeed locally located, if not entirely locally produced. The intimate connections between the daily lives of women and the life of the city suggest that political activity by coalitions of women aimed specifically at defining inclusive, "good" communities might result in localities that are less marginalizing, hierarchical, and dangerous. Far from presenting an inherent obstacle to the attainment of feminist goals or of diversity, then, informal and formal local communities are potential sources of women's empowerment and political change.

Conceptual wrangling will not resolve the tensions around the continuing, but highly problematic, relevance of place-based community for many women. At the same time, abandoning the name *community* for fear of invoking unsavory political arrangements lurking beneath it will not relieve us of the two basic "problems" that people do have different relationships with community and that the idea of local community is, apparently, much less appealing to feminist theorists than to women in urban contexts. Even supposing it is possible, genuinely, to accommodate the demands of more women to be citizens of open and humane cities, it remains to be demonstrated which communities are the best vehicles for that transformation. It has not been argued persuasively that local community is unworthy of feminist attention, and the "community problem" awaits further examination.

REFERENCES

Ackelsberg, M. (1988). Communities, resistance, and women's activism: Some implications for a democratic polity. In A. Bookman & S. Morgen (Eds.), *Women and the politics of empowerment* (pp. 297-313). Philadelphia: Temple University Press.

Anderson, J. (1994). Separatism, feminism, and the betrayal of reform. *Signs, 19*, 437-447.

Andrew, C. (1992). The feminist city. In H. Lustiger-Thaler (Ed.), *Political arrangements: Power and the city* (pp. 109-122). Montréal: Black Rose Books.

Barber, B. (1984). *Strong democracy: Participatory politics for a new age.* Berkeley: University of California Press.

Beauregard, R. A. (1993). Descendants of ascendant cities and other dualities. *Journal of Urban Affairs, 15,* 217-229.

Bell, D., & Bansal, P. (1988). The republican revival and racial politics. *Yale Law Journal, 97,* 1609-1622.

Bellah, R. N., Madsen, R., Sullivan, W. M., Swidler, A., & Tipton, S. M. (1985). *Habits of the heart: Individualism and commitment in American life.* Berkeley: University of California Press.

Bookman, A., & Morgen, S. (Eds.). (1988). *Women and the politics of empowerment.* Philadelphia: Temple University Press.

Boyte, H. C. (1984). *Community is possible: Repairing America's roots.* New York: Harper & Row.

Butler, J., & Scott, J. W. (Eds.). (1992). *Feminists theorize the political.* New York: Routledge.

Cornell, D. (1987). Two lectures on the normative dimensions of community in the law. *Tennessee Law Review, 54,* 327-343.

Dietz, M. (1985). Citizens with a maternal face: The problem with maternal thinking. *Political Theory, 12,* 19-37.

Fowler, R. B. (1991). *The dance with community.* Lawrence: University Press of Kansas.

Friedman, M. (1989). Feminism and modern friendship: Dislocating the community. *Ethics, 99,* 275-290.

Gilkes, C. T. (1988). Building in many places: Multiple commitments and ideologies in Black women's community work. In A. Bookman & S. Morgen (Eds.), *Women and the politics of empowerment* (pp. 53-76). Philadelphia: Temple University Press.

Gittell, M., & Shtob, T. (1980). Changing women's roles in political volunteerism and reform of the city. *Signs, 5,* S67-S78.

Greene, M. F. (1991). *Praying for sheetrock.* New York: Addison-Wesley.

Gutmann, A. (1985). Communitarian critics of liberalism. *Philosophy and Public Affairs, 14,* 308-322.

Haeberle, S. H. (1993, September). *Cliques, crowds, and clubs: A comparative look at gays and lesbians as interest groups in urban politics.* Paper presented at the annual meeting of the American Political Science Association, Washington, DC.

Hekman, S. (1992). The embodiment of the subject: Feminism and the communitarian critique of liberalism. *Western Politics Quarterly, 54,* 1098-1119.

Hoch, C., & Slayton, R. A. (1989). *New homeless and old: Community and the skid row hotel.* Philadelphia: Temple University Press.

Hummon, D. M. (1990). *Commonplaces: Community ideology and identity in American culture.* Albany: State University of New York Press.

Jacobs, J. (1961). *The death and life of great American cities.* New York: Random House.

Luttrell, W. (1988). The Edison School struggle: The reshaping of working-class education and women's consciousness. In A. Bookman & S. Morgen (Eds.), *Women and the politics of empowerment* (pp. 136-156). Philadelphia: Temple University Press.

MacIntyre, A. (1981). *After virtue.* South Bend, IN: University of Notre Dame Press.

Mansbridge, J. (1980). *Beyond adversarial democracy.* New York: Basic Books.

Martinez-Brawley, E. E. (1990). *Perspectives on the small community.* Silver Spring, MD: National Association of Social Workers Press.

Michelman, F. (1988). Law's republic. *Yale Law Journal, 97,* 1493-1538.

Morgen, S. (1988). "It's the whole power of the city against us!": The development of political consciousness in a women's health care coalition. In A. Bookman & S. Morgen (Eds.), *Women and the politics of empowerment* (pp. 97-115). Philadelphia: Temple University Press.

Mouffe, C. (1991). Democratic citizenship and the political community. In the Miami Theory Collective (Eds.), *Community at loose ends* (pp. 70-82). Minneapolis: University of Minnesota Press.

Mouffe, C. (1992). Feminism, citizenship and radical democratic politics. In J. Butler & J. W. Scott (Eds.), *Feminists theorize the political* (pp. 369-384). New York: Routledge.

Nicholson, L. J. (Ed.). (1990). *Feminism/postmodernism.* New York: Routledge.

Pateman, C. (1988). *The sexual contract.* Berkeley: University of California Press.

Pateman, C. (1989). *The disorder of women: Democracy, feminism, and political theory.* Cambridge, UK: Polity.

Phelan, S. (1989). *Identity politics: Lesbian feminism and the limits of community.* Philadelphia: Temple University Press.

Rayside, D. (1989). Small town fragmentation and the politics of community. *Journal of Canadian Studies, 24,* 103-120.

Rose, D., & Villeneuve, P., with F. Colgan. (1988). Women workers and the inner city: Some implications of labor force restructuring in Montréal, 1971-81. In C. Andrew & B. Moore Milroy (Eds.), *Life spaces: Gender, household, and employment* (pp. 31-64). Vancouver: University of British Columbia Press.

Saegert, S. (1988). The androgenous city: From critique to practice. In W. Van Vliet (Ed.), *Women, housing, and community* (pp. 23-37). Aldershot, UK: Avebury.

Sandel, M. (1982). *Liberalism and the limits of justice.* Cambridge, UK: Cambridge University Press.

Sherry, S. (1986). Civic virtue and the feminine voice in constitutional adjudication. *Virginia Law Review, 72,* 543-616.

Shilts, R. (1988). *And the band played on: People, politics, and the AIDS epidemic.* New York: Penguin.

Shlay, A. B. (1986). Women, space, and community: A feminist agenda. *Urban Resources, 3,* Ba1-Ba4.

Sullivan, K. M. (1988). Rainbow republicanism. *Yale Law Journal, 97,* 1713-1723.

Sunstein, C. (1988). Beyond the republican revival. *Yale Law Journal, 97,* 1539-1590.

Sypnowich, C. (1993). Justice, community, and the antinomies of feminist theory. *Political Theory, 21,* 484-506.

Taylor, C. (1985). Alternative futures: Legitimacy, identity and alienation in late twentieth century Canada. In A. Cairns & C. Williams (Eds.), *Constitutionalism, citizenship and society in Canada* (pp. 183-229). Toronto: University of Toronto Press.

Wekerle, G. (1988a). Canadian women's housing cooperatives: Case studies in physical and social innovation, In C. Andrew & B. Moore Milroy (Eds.), *Life spaces: Gender, household, and employment* (pp. 102-140). Vancouver: University of British Columbia Press.

Wekerle, G. (1988b). From refuge to service center: Neighborhoods that support women. In W. Van Vliet (Ed.), *Women, housing, and community* (pp. 23-37). Aldershot, UK: Avebury.

Wilder, M. G. (1993). Institutional processes: Shaping place within disciplinary walls, a reply to Anne B. Shlay. *Journal of Urban Affairs, 15,* 405-411.

Young, I. M. (1989). Polity and group difference: A critique of the ideal of universal citizenship. *Ethics, 99,* 250-274.

Young, I. M. (1990a). The ideal of the community and the politics of difference. In L. J. Nicholson (Ed.), *Feminism/postmodernism* (pp. 300-323). New York: Routledge.

Young, I. M. (1990b). *Justice and the politics of difference.* Princeton, NJ: Princeton University Press.

3   The Gender Regimes of American Cities

LYNN M. APPLETON

All cities are patriarchal, but neither all cities nor all patriarchies are the same. Although all patriarchies are systems "of social structures, and practices in which men dominate, oppress and exploit women" (Walby, 1989, p. 214), they differ in their structure, conflicts, and discontents. Cities differ in their "gender regimes" (Connell, 1987): the ways in which their political, economic, and familial systems combine to produce gender inequality. A distinctive gender regime is supported by the low-density, relatively homogeneous style of development in suburban America, prevalent in post-1960s urban growth. Inside this kind of urban place, lives are still shaped by the 19th century's ideology of "separate spheres" (Bose, 1987) for men and women. Struggles over gender are privatized, interpersonal, and intrapsychic. Higher density and more heterogeneous central cities produce a very different gender regime than do suburbs. In these cities, the contradictions of the separate spheres have been heightened and forced into increasingly public and collective realms. Struggles over gender have spilled over into political and workplace struggles, as well as into neighborhood and "lifestyle" politics.

This chapter explores the relationship between variation in urban form and gender relations, drawing on recent work in feminist theory that stresses that there are variations in patriarchal structures. Early feminist work often characterized patriarchy as monolithic and invariant, as if male dominance was the same in medieval France and contemporary Chicago. But feminist scholars recognize that patriarchy varies substantially across time and place as women and "lesser men" respond to the initiatives of privileged men. Not all women are equally oppressed, nor are all men equally oppressors. Degree of male control of women's lives varies substantially (see Kandiyoti, 1988).

Newer work on patriarchy conceptualizes it as a loosely coupled system that produces gender inequality and that is shaped by men's and women's interests, resources, and actions as they negotiate and struggle over power and authority. By incorporating feminist theory on patriarchy into urban studies, we can understand an additional dimension of urban life and politics, including the complex ways in which cities are sites for both reproducing and challenging patriarchy. This research will aid in developing an understanding of the tremendous variation in the ways in which cities maintain the stability of the dominant patriarchal order.

■ Patriarchal Institutions

To understand the gendered consequences of urban form, the city must be conceptualized as the nexus of three basic institutions that shape patriarchy: the family, the economy, and the state.

The family is a primary site for enacting gender and thereby maintaining patriarchy. The heterosexual family is a strongly gendered institution where only husbands and wives exist, not spouses, and where only mothers and fathers exist, not parents. This family form is not culturally universal. Rather, socially structured resources and opportunities explain men's and women's motivations for and roles in marriage (see Epstein, 1988; Hochschild, 1989), as well as the role of marriage and childbearing in reproducing sex inequality.

Marriage and parenthood have dramatically different consequences for men and women. They contribute to gender inequality because they place a disproportionate burden on women as spouses and parents. These gendered burdens are not solely a consequence of female economic dependence on men. Although individual women's economic resources do have some effect on individual men's participation in housework (see Coleman, 1991), socialized ideas of "women's place" are a more important determinant of the family's division of labor (Cancian, 1987). Regardless of which spouse makes more money, wives consistently carry the greatest burden of child care, household chores, and neighboring activities (Calansanti & Bailey, 1991; Campbell & Lee, 1990; LaRossa & LaRossa, 1989).

Thus research on the family as a gendered and patriarchal institution argues that it is a primary site for our feelings about gender's importance and our ideas about its meaning, an everyday site for the enactment of gender, and an important site of reproduction of sex inequality.

The economy is also a primary site for enacting gender and constructing gender relations. Historically, an increase in gender inequality has been associated with women's exclusion from work that produces exchange value and is performed away from the home. Such exclusion is associated with the assignment of women to the devalued and often invisible work of "secondary production" that is child care, cooking, and cleaning (O'Kelly & Carney, 1986).

In studying industrial societies, research on gendered work focuses on the causes, dynamics, and consequences of women's exclusion from waged work or segregated access to it (Blumberg, 1991; England & McCreary, 1987). When women have paying jobs, their jobs differ from men's in that they are less various, more poorly paid, and offer less opportunity (see Ferree, 1987).

But research on capitalist economies reveals cross-national variation in occupational segregation (see Charles, 1992) and women's average wages (Norris, 1987). These findings suggest that state differences are significant in shaping gender inequality.

Two general theories of the state appear in feminist work (Connell, 1990). One argues that the state is the tool of a patriarchy anchored in family or economy. The other argues that the state itself is patriarchal. The former argument underpins liberal feminism's argument that the state can be captured by feminist interests, whereas the latter suggests that the state is as intractably patriarchal as family or the economy.

Some researchers focus on barriers to women's political activity (Gelb & Gittell, 1986; Matthews, 1992) or on effects of electoral and party systems on gender equality (Norris, 1987). But the largest group of research focuses on how ostensibly "gender neutral" state policies and practices produce gendered outcomes that reproduce sex inequality (see Diamond, 1983).

■ Institutional Intersections: The City

Urban scholars are accustomed to conceptualizing cities as political and economic systems. But to understand how cities are part of systems of sex stratification, family systems must be included in urban analysis. Each city has a distinctive relationship between its political, economic, and familial systems that constructs its gender regime, its particular version of patriarchy.

I have taken the concept of *gender regime* from the work of theorists who have reconceptualized gender as the product of an ongoing series of struggles within and between social institutions (Connell, 1987). A gender regime is the way that gender is shaped by and shapes a particular social institution or, in the case of the city, a confluence of social institutions. Each city has a gender regime that it shares with similarly constituted cities. Urban gender regimes can be characterized in terms of the prevailing ideologies of how men and women should act, think, and feel, the availability of cultural and behavioral alternatives to those ideologies, men's and women's access to social positions and control of resources, and the relationships between men and women.

Different kinds of cities have different kinds of gender regimes. For example, contemporary New York City has a gender regime that provides an alternative to male/female emotional and economic interdependency. Extensive subcultures of unmarried persons, both heterosexual and homosexual, provide an alternative to heterosexual marriage. Individualist assumptions about adults' lives are a significant counterweight to the familism that is still the dominant American ideology of adulthood.

Such deviant ways of life flourish only in the density and heterogeneity of the urban place (Abrahamson & Sigelman, 1987). Consider New York City's neighbors, the less dense and more homogeneous suburban communities of Long Island. The gender regimes of these cities provide fewer alternatives to heterosexual marriage. The mature adult is expected to be married, and the single-family home is assumed to be an adult's economic goal. No cultural counterweight exists to what Rich (1980) characterizes as "compulsory heterosexuality." These cities have no social space for an oppositional gay subculture (D'Emilio, 1983) or a subculture of female economic independence. In these cities, all adults are expected to marry, and women's primary responsibilities are those of home and family, whereas men's responsibilities are primarily economic.

These two kinds of cities have very different gender regimes: private patriarchy in the Long Island suburbs and a more public patriarchy in the New York central city. Private patriarchy is that set of gender regimes premised on a strongly gendered division of labor inside the home, paid work segregated by and remunerated on the basis of workers' sex, and women's dependence on the income of individual men (Brown, 1987). The main site of gender struggle is the family, as spouses contest over scarce resources such as time and money. The hegemony of heterosexuality, parenthood, and marriage are unquestioned.

Public patriarchy is that set of gender regimes premised on women's increasing economic independence of individual men, increasing dependence on the state for income, and decreasing emotional interdependence with men. Paid work continues to be segregated by and remunerated on the basis of workers' gender, but it becomes more central to women's lives. The hegemony of heterosexuality, maternity, and motherhood is challenged when patriarchal ideologies are strained as alternate visions of men's and women's lives develop. Although gender struggle continues inside heterosexual relationships, it increasingly moves to the public realm of women's struggles for better paid jobs, state assistance, and public policies that serve their interests.

The remainder of this chapter will outline the relationship between urban form and these two patriarchal forms. First, I will discuss how the current gender regimes of the central cities were shaped by the urban changes of the 1950s and 1960s. Next, I will discuss the preservation of the private patriarchy in the suburbs of the 1950s and the "uncentered" cities of the most recent spurt of urban growth.

■ The Public Patriarchy of the Central Cities

Currently, the private patriarchy of the suburbs and uncentered cities is the dominant American patriarchal form: privatized struggles between men and women and a strong culture of separate spheres that define masculinity and femininity (Bose, 1987). In the early 20th century, this was also the patriarchal form that had emerged from the gender struggles of earlier decades (Piess, 1986; Stansell, 1986). But beginning in the 1960s, the central cities have produced a different kind of patriarchy. This "public patriarchy" (Brown, 1987) is characterized by multiple masculinities and femininities, open internal dissension in the gender realm, and increasingly public and collective struggles over men's and women's access to resources. It emerged as the composition of the central cities increasingly contrasted with that of the suburbs of the 1950s and, later, of the new cities that emerged from the urban growth in the 1970s and 1980s.

First, because of its demographic composition, the central city contains a disproportionate number of people predisposed to challenge gender inequality. The majority (73% in 1984) of all female householders are central-city dwellers (Cook & Rudd, 1984), and women who live in households without a male wage earner are the most likely to be aware of and want to see change in gender inequality (Davis & Robinson, 1991).

Second, in the 1960s, the leading activists of the second wave of the feminist movement were drawn from the educated elites of these cities (Connell, 1987, p. 271). Third, the central cities produced important challenges to patriarchy from sources that often were unmotivated by conscious criticism of gender inequality, such as the proportion of central-city families that had two wage earners, the proportion of urban women who were single mothers, and the significant proportion of urban dwellers who lived independently of marriage and children. The influence of these groups increased during the 1960s, as industry and more affluent families abandoned the central cities. At this time, patriarchy's most significant "boosters" increasingly abandoned the city to groups that were more critical of the existing system of sex stratification.

Dual-Wage-Earner Families

Despite the prescriptions of the "cult of domesticity" (Bose, 1987), women in the central city of this century were never as much the captives of their homes as were the women of suburbia (Saegert, 1981; Wekerle, 1981). First, even though their income was considered a lesser and "second" income, many urban married women had paying jobs. Although workforce participation varied across decades, city women were more apt to be in the paid workforce than were nonurban women. They found it easier to get work compatible with domestic responsibilities: close to home, part-time, and not requiring additional family investment (e.g., in a second car). In a setting in which many married women worked, spouses found it easier to define wives' paid work as legitimate and "normal." In particular, African American and other minority women found it easy to justify their paid jobs as an essential contribution to family welfare in a racist society (Collins, 1991; Glenn, 1991).

Second, urban density and the structure of public space provided city women with more interactional opportunities than those of suburban women. They had access to extensive networks of family members and, to a lesser extent, women friends. Consequently, their identities were less embedded in husband, children, and home than were those of suburban women.

Third, urban life provided a level of stimulation and challenge that distinguished the urban woman from her suburban counterpart. Thus the central city constructed a more independent wife than did the suburb. Fourth, city homes were smaller and more of their work could be contracted out. City women's housekeeping was a lesser focus of their lives.

Finally, urban husbands were not as dependent on their families for their social and emotional life as were suburban husbands. In the city, same-sex sociability was a normal feature of adult men's lives. Consequently, many urban families differed from private patriarchy's early prescriptions and from the emerging reality of American suburban life.

Single Mothers

In any economy characterized by occupational sex segregation and lower wages for women, most husbandless women are poor. So in the 1960s and 1970s, rates of women's poverty increased as rates of divorce and nonmarital childbearing increased. As McLanahan, Sorensen, and Watson (1989) point out, the increase in women's poverty rates was due mostly to women's responsibility for children. Both men and women became increasingly likely to abandon or not assume spousal roles, but whereas men abandon fatherhood with alacrity, women cling to motherhood.

As the number of husbandless mothers increased, their public benefits declined. Benefit levels failed to keep pace with inflation and more recently, have been directly cut. Current political discourse is dominated by talk about "workfare" and "job training." The clear implication is that mothers should find paid work to support their children. But because nearly half of working women are employed in industries that pay poverty-level wages for full-time work (McLanahan et al., 1989, p. 120), such solutions are unrealistic.

Because most female-headed households are urban households, the feminization of poverty might be a central issue for urban politics. But husbandless women do not mobilize or gain power easily (see West, 1981). Hence their problems are seen as private problems that should be solved by finding a husband or a job.

Unmarried Adults, Often Without Children

In addition to dual-wage-earner families and single mothers, the central city contrasts with the suburbs in its higher proportion of young, single adults. As people married later in life, young men and women worked and lived independently for longer periods of time.

The numbers of women living independently of families were historically unprecedented. They controlled money, made decisions, and negotiated with individual men over the course of a relationship. They were neither daughters, ladies, nor hardworking wives (Piess, 1986). As urban dwellers, they had a distinctive set of preferences for their neighborhoods (Shlay & DiGregorio, 1985). Their emerging subculture subverted the old patriarchal order.

Gay men and lesbians were a second source of challenge to the patriarchal order. During the 20 century, an increasingly large and well-organized gay community developed in major American cities (D'Emilio & Freedman, 1988). By the 1960s, gay men had begun to carve out neighborhoods for themselves, develop elaborate social and economic institutions, and create the basis for a relatively self-contained gay subculture. By mid-century, lesbian communities also emerged in the larger American cities (Faderman, 1991). But trapped in a world of poorly paid women's work, lesbians' limited economic power lessened their ability to create neighborhoods on the scale of those created by gay men.

Initially, lesbians and gay men were a small challenge to private patriarchy. But as the liberation movements of the late 1960s and early 1970s developed, post-Stonewall homosexual communities were an increasingly visible feature of the central city. They challenged assumptions about masculinity and femininity and contributed to the central city's increasingly open debates about the future of patriarchy.

■ From City to Suburb: Private Patriarchy Preserved

Private patriarchy emerged in the cities of the industrial era, but it found its highest expression in the suburbs of the 1950s and 1960s. In the 1970s and later, it was institutionalized in the new cities of autonomous suburban belts and sunbelt sprawls. It survived in the landscape of the American Dream: a setting of single-family homes, physically separated commercial and residential activities, social class homogeneity, and racial exclusiveness.

Separate Spheres in the Suburbs

Initially, the gender regime of the suburbs was a private patriarchy with a strictly gendered division of labor linked to the heterosexual nuclear family. Men were expected to do paid work, earn a family wage, and support women and children. Women were assigned the unpaid work of social reproduction: "the activities and attitudes, behaviors and emotions, responsibilities and relationships directly involved in the maintenance of a life on a daily basis, and intergenerationally" (Laslett & Brenner, 1989, pp. 382-383). This gender regime had been developed in the early part of the 20th century, survived the challenges of women's changing roles during World War II, and was part of the idealized family sought by the new suburbanites.

Men first embraced the "breadwinner/homemaker" bargain during the latter part of the 19th century (see Laslett & Brenner, 1989). But their ability to achieve breadwinner status required access to the "family wage." Although some men attained this wage through class position, working- and middle-class men's access to it required the successes of the union movement, the protective labor legislation of the 1930s, and a set of national political and economic changes that greatly improved their economic status.[1] The economic prosperity of the 1950s, combined with federal subsidies for home mortgages, made the suburban exodus possible and brought the breadwinner role within reach for a significant proportion of European American men.

Women began their embrace of the "homemaker" role in the late 19th century, although its realization rested on the later economic changes that ensured the family wage for their husbands (Cancian, 1987; Geschwender, 1992; Laslett & Brenner, 1989). They were both pushed and pulled into a primary investment in the work of social reproduction. Child care and housework were less compatible with industrial, paid jobs than with the household production of preindustrial societies. Additionally, "women's jobs" in the industrial economy were generally poorly paid and without intrinsic reward. But the valorization of the separate spheres also has been analyzed as part of women's struggle to justify their withdrawal from waged work, to legitimate a claim on men's income, and to raise their status by creating a new idea of "women's place" (Bose, 1987; Cancian, 1987). By the 1950s, the homemaker role was widely embraced by women migrants to the suburbs (Saegert, 1981). In the suburbs, these relatively affluent and overwhelmingly White women hoped to live out the prescription that "a woman's place is in the home."

Women's Work in the Household

For cultural and practical reasons, suburban houses and yards required more work to decorate, clean, and maintain than did urban rental apartments. They were larger, more elaborate, and prized as a family's only investment. Moreover, by the 1950s, a woman's homemaking was seen as a way to express love for her family and to define her individuality (Cowan, 1983; Laslett & Brenner, 1989). Additionally, although the new "consumerist" ethic rapidly defined women's roles as purchasers of household goods, shopping retreated to "shopping centers" distant from residential neighborhoods. But because of women's lesser access to cars (Salem, 1986, p. 155), shopping was more time-consuming and complicated than in an urban setting.

Even if a woman had been willing to hire household labor, such labor was scarce in the suburbs. The absence of nearby poor women made domestic labor difficult to hire. Even today, poor minority women travel from their urban homes to the suburban ring to do domestic work (Hwang & Fitzpatrick, 1992), but the lack of public transportation makes such commuting difficult. Further, the low density of suburban communities lessened the profitability of businesses that might have decreased household labor (e.g., laundries).

Child care became more labor-intensive in the suburbs. The suburban exodus was at the height of the new child-centered culture of child rearing (Rossi, 1987). Repeatedly, people who moved to the suburbs stressed that they were seeking "a good place to bring up children" (Popenoe, 1977; Saegert, 1981). But low-density neighborhoods and the absence of public transportation limited the pool of potential activities and playmates for children. Because the suburbs provided a range of specialized activities for children (e.g., Little League, dance lessons), "Mom's taxi" emerged as a way of life.

Thus the suburbs segregated men's and women's lives and supported a strongly gendered division of labor.[2] For a short period in American urban history, a new kind of purdah appeared: married women isolated and controlled in the mostly female setting of daytime suburbia while married men ventured into the profane world of work and danger in the central cities. But despite Friedan's (1963) observations in *The Feminine Mystique*, suburban women generally concluded that their lost opportunities for adult interaction and paid work were compensated by their children's and husband's greater pleasure in the new way of life (Popenoe, 1977; Saegert, 1981; Wilson, 1991).

Women's Paid Work

Initially, suburban women's economic dependence on men was ensured by the ideology of the cult of domesticity, adequate husbandly income, and the absence of suburban economic opportunities for women. But the growth of suburban belts in the 1960s soon shifted patterns of industrial and commercial location. As suburbs attracted men's jobs, they also drew complementary jobs in office work. At first, these were reserved for "single girls." But, as these jobs multiplied, two factors changed women's competition for them: a shortage of single girls and a change in norms for working wives and mothers.

Norms for married women's work changed in response to global economic restructuring. In the early 1970s, the income of the average American worker stagnated and then declined. But American families' material desires did neither so, increasingly, married women's paid work was justified as a way of maintaining the family's standard of living (Geschwender, 1992). When they worked, women adopted strategies to define their jobs as "helping out" their husbands rather than as challenges to the separate spheres ideology (Rosen, 1987). Married women with school-age children increasingly sought paid work (Bose, 1987) and, by the late 1980s, about half of the mothers of children under 3 years of age held paying jobs (Ferree, 1987). The change in workforce participation was greatest for White women, because the poverty of minority Americans had long required a dual-wage-earner family.

Although women's and men's jobs differ in the city as well as in the suburbs, women may have less access to good jobs in the suburban setting. First, many of the better paying "women's jobs" are located in the organizational headquarters and government offices of the central city. But because of their household responsibilities, women generally restrict their search for work to jobs near their homes. In low-density settlements that separate residential and commercial areas, opportunities for women's work are severely restricted (Hwang & Fitzpatrick, 1992; Popenoe, 1977; Semyonov & Lewin-Epstein, 1991). Thus suburban working women are unable to close the wage gap in their families.

Paid Work and Household Work

As suburban married women moved into the workforce, new tensions appeared in the private patriarchy. Wives' paid work produced tremendous strain in households in which women still retained almost sole responsibility for child care and housework. Hochschild (1989) estimates that working the "double shift" requires American women with children to work a month per year more than their husbands do. An increasing proportion of American women report that they are overtired, overworked, and overwhelmed by their responsibilities at work and at home (Schor, 1991). Wives' increased earnings and hours of paid work rarely produce comparable increases in husbands' household work (Berk, 1985). In tandem with rising standards of marital happiness, the unequal household burdens of husband and wife have produced high levels of conflict and discontent in American families.

Divorce

Women's marital discontent is likely to increase as their involvement in paid work increases (Booth, Johnson, White, & Edwards, 1984; Hochschild, 1989). When they are discontented, however, women tend to believe that their marital problems are caused by their husbands rather than by external, structural factors. Divorce has become an increasingly frequent solution to marital problems but also a new source of gender inequality. Divorced women's low incomes and primary responsibility for children's support further disadvantages them relative to men (Goldberg, 1990; Weitzman, 1985). Generally, they are poor (McLanahan et al., 1989), and the suburbs have few places for them to live and few jobs that permit them to support their families.

Female-Headed Households in the Suburbs

The suburbs and low-density cities assume that a married couple is the normal household. Newly divorced women supporting children have trouble finding a job that pays a family wage and affordable rental property that permits children. Despite their increased need for the support and shared resources of a network of friends or kin, suburban single mothers are likely to be more socially isolated than mothers in an urban setting.

In denser and more heterogeneous central cities, the range of housing options is greater. More and better jobs are available to women. Public transportation and a wide range of social services are available, and urban women are better able to live near networks of friends and kin. Additionally, poor urban women with children have a greater likelihood of seeking public or collective solutions to their problems than do poor suburban women. The tradition of stronger state presence and collective action in the cities stands in stark contrast to the tradition of minimal government in the suburbs and uncentered cities.

■ Conclusion

Although most Americans live in urbanized areas, most do not live in densely settled and heterogeneous central cities. Most live in low-density and relatively homogeneous urban places that are the linear descendants of the first suburban explosion. Therefore, most live in settings that support a private patriarchy that originated in but migrated from the central city. And if trends in American urbanization continue unabated, the dominance of the private patriarchy is likely to continue.

But in this chapter, I have argued that the dominance of private patriarchy is neither assured nor uniform. Cities vary in their gender regimes, with consequences for the degree and legitimacy of sex stratification. For example, cities vary in their zoning regulations and support for public transportation, and their policies have significant effects on women's access to paid work. Cities vary in their proportion of unmarried adults, and this variation affects their residents' perceptions of what constitutes "normal" adult life. Interurban variation in gender relations is as significant as the ways in which cities differ in their class composition or system of racial/ethnic relations. The higher density and more heterogeneous central cities offer challenges to the dominant system of sex stratification that are absent in the gender regimes of lower density and more homogeneous urban places. An understanding of changing patterns of gender struggle requires an understanding of urban change and, because urban change has been shaped by gendered ideals of personal and family life, urban analysis requires an analysis of gender struggles.

NOTES

1. In this, I include a range of phenomena: in the 1920s, the dramatic decrease in immigration; the American post-World War II economic boom; the federal government's willingness to subsidize private home ownership through low mortgages and tax deductions. Note, however, that these changes disproportionately benefited White men and excluded men of color.

2. Nancy Chodorow (1978), leading theorist of the sociological psychoanalytic approach to gender, argues that contemporary American masculine and feminine character are consequences of infancy in the gendered world of separate spheres, characteristic of the suburbs.

REFERENCES

Abrahamson, M., & Sigelman, L. (1987). Occupational sex segregation in metropolitan areas. *American Sociological Review, 52,* 588-597.

Berk, S. F. (1985). *The gender factory: The apportionment of work in American households.* New York: Plenum.

Blumberg, R. L. (1991). Introduction: The "triple overlap" of gender stratification, economy, and the family. In R. L. Blumberg (Ed.), *Gender, family, and economy: The triple overlap* (pp. 7-34). Newbury Park, CA: Sage.

Booth, A., Johnson, D. R., White, L., & Edwards, J. N. (1984). Women, outside employment, and marital instability. *American Journal of Sociology, 90,* 567-583.

Bose, C. (1987). Dual spheres. In B. B. Hess & M. M. Ferree (Eds.), *Analyzing gender: A handbook of social science research* (pp. 267-285). Newbury Park, CA: Sage.

Brown, C. (1987). The new patriarchy. In C. Bose, R. Feldberg, & N. Sokoloff (Eds.), *Hidden aspects of women's work* (pp. 137-159). New York: Praeger.

Calansanti, T. M., & Bailey, C. A. (1991). Gender inequality and the division of household labor in the United States and Sweden: A socialist-feminist approach. *Social Problems, 38,* 34-53.

Campbell, K., & Lee, B. A. (1990). Gender differences in urban neighboring. *Sociological Quarterly, 31,* 495-512.

Cancian, F. (1987). *Love in America: Gender and self-development.* New York: Cambridge University Press.

Charles, M. (1992). Cross-national variation in occupational sex segregation. *American Journal of Sociology, 57,* 483-502.

Chodorow, N. (1978). *The reproduction of mothering.* Berkeley: University of California Press.

Coleman, M. T. (1991). The division of household labor: Suggestions for future empirical consideration and theoretical development. In R. L. Blumberg (Ed.), *Gender, family, and economy: The triple overlap* (pp. 245-260). Newbury Park, CA: Sage.

Collins, P. H. (1991). *Black feminist thought: Knowledge, consciousness, and the politics of empowerment.* New York: Routledge.

Connell, R. W. (1987). *Gender and power: Society, the person and sexual politics.* Stanford, CA: Stanford University Press.

Connell, R. W. (1990). The state, gender, and sexual politics: Theory and appraisal. *Theory and Society, 19,* 507-544.

Cook, C., & Rudd, N. M. (1984). Factors influencing the residential location of female householders. *Urban Affairs Quarterly, 20,* 78-96.

Cowan, R. S. (1983). *More work for mother: The ironies of household technology from the open hearth to the microwave.* New York: Basic Books.

Davis, N. J., & Robinson, R. V. (1991). Men's and women's consciousness of gender inequality: Austria, West Germany, Great Britain, and the United States. *American Sociological Review, 56,* 72-84.

D'Emilio, J. (1983). Capitalism and gay identity. In A. Snitow, C. Stansell, & S. Thompson (Eds.), *Powers of desire: The politics of sexuality* (pp. 100-116). New York: Monthly Review Press.

D'Emilio, J., & Freedman, E. B. (1988). *Intimate matters: A history of sexuality in America.* New York: Harper & Row.

Diamond, I. (Ed.). (1983). *Families, politics, and public policy: A feminist dialogue on women and the state*. New York: Longman.

England, P., & McCreary, L. (1987). Gender inequality in paid employment. In B. B. Hess & M. M. Ferree (Eds.), *Analyzing gender: A handbook of social science research* (pp. 286-321). Newbury Park, CA: Sage.

Epstein, C. F. (1988). *Deceptive distinctions: Sex, gender, and the social order*. New Haven, CT: Yale University Press.

Faderman, L. (1991). *Odd girls and twilight lovers: A history of lesbian life in twentieth-century America*. New York: Penguin.

Ferree, M. M. (1987). She works hard for a living: Gender and class on the job. In B. B. Hess & M. M. Ferree (Eds.), *Analyzing gender: A handbook of social science research* (pp. 322-347). Newbury Park, CA: Sage.

Friedan, B. (1963). *The feminine mystique*. New York: Dell.

Gelb, J., & Gittell, M. (1986). The role of activist women in cities. In J. Boles (Ed.), *The egalitarian city: Issues of rights, distribution, access, and power* (pp. 93-109). New York: Praeger.

Geschwender, J. A. (1992). Ethgender, women's waged labor, and economic mobility. *Social Problems, 39*, 1-15.

Glenn, E. N. (1991). Racial ethnic women's labor: The intersection of race, gender, and class oppression. In R. L. Blumberg (Ed.), *Gender, family, and economy: The triple overlap* (pp. 173-200). Newbury Park, CA: Sage.

Goldberg, G. S. (1990). The United States: Feminization of poverty amidst plenty. In G. S. Goldberg & E. Kremen (Eds.), *The feminization of poverty: Only in America?* (pp. 17-58). New York: Greenwood.

Hochschild, A. R. (1989). *The second shift*. New York: Avon.

Hwang, S., & Fitzpatrick, K. M. (1992). The effect of occupational sex segregation and the spatial distribution of jobs on commuting patterns. *Social Science Quarterly, 73*, 550-564.

Kandiyoti, D. (1988). Bargaining with patriarchy. *Gender and Society, 2*, 274-290.

LaRossa, R., & LaRossa, M. M. (1989). Baby care: Fathers vs. mothers. In B. J. Risman & P. Schwartz (Eds.), *Gender in intimate relationships: A microstructural approach* (pp. 138-154). Belmont, CA: Wadsworth.

Laslett, B., & Brenner, J. (1989). Gender and social reproduction: Historical perspectives. *Annual Review of Sociology, 15*, 381-404.

Matthews, G. (1992). *The rise of public woman: Woman's power and woman's place in the United States, 1630-1970*. New York: Oxford University Press.

McLanahan, S. S., Sorensen, A., & Watson, D. (1989). Sex differences in poverty, 1950-1980. *Signs, 15*, 102-122.

Norris, P. (1987). *Politics and sexual equality: The comparative position of women in Western democracies*. Boulder, CO: Rienner.

O'Kelly, C. G., & Carney, L. S. (1986). *Women and men in society: Cross-cultural perspectives on gender stratification*. Belmont, CA: Wadsworth.

Piess, K. (1986). *Cheap amusements: Working women and leisure in turn-of-the-century New York*. Philadelphia, PA: Temple University Press.

Popenoe, D. (1977). *The suburban environment: Sweden and the United States*. Chicago: University of Chicago Press.

Rich, A. (1980). Compulsory heterosexuality and the lesbian continuum. *Signs, 5*, 631-660.

Rosen, E. I. (1987). *Bitter choices: Blue-collar women in and out of work.* Chicago: University of Chicago Press.

Rossi, A. S. (1987). Parenthood in transition: From lineage to child to self-orientation. In J. B. Lancaster, J. Altmann, A. S. Rossi, & L. R. Sherrod (Eds.), *Parenting across the life span: The biosocial dimension* (pp. 31-84). New York: Aldine de Gruyter.

Saegert, S. (1981). Masculine cities and feminine suburbs: Polarized ideas, contradictory realities. In C. R. Stimpson, E. Dixler, M. J. Nelson, & K. B. Yatrakis (Eds.), *Women and the American city* (pp. 93-108). Chicago: University of Chicago Press.

Salem, G. (1986). Gender equity and the urban environment. In J. Boles (Ed.), *The egalitarian city: Issues of rights, distribution, access, and power* (pp. 152-161). New York: Praeger.

Schor, J. B. (1991). *The overworked American: The unexpected decline of leisure.* New York: Basic Books.

Semyonov, M., & Lewin-Epstein, N. (1991). Suburban labor markets, urban labor markets, and gender inequality in earning. *Sociological Quarterly, 32,* 611-620.

Shlay, A. B., & DiGregorio, D. A. (1985). Same city, different worlds: Examining gender- and work-based differences in perceptions of neighborhood desirability. *Urban Affairs Quarterly, 21,* 66-86.

Stansell, C. (1986). *City of women: Sex and class in New York, 1789-1860.* New York: Alfred A. Knopf.

Walby, S. (1989). Theorizing patriarchy. *Sociology, 23,* 213-234.

Weitzman, L. (1985). *The divorce revolution: The unexpected consequences for women and children.* New York: Free Press.

Wekerle, G. R. (1981). Women in the urban environment. In C. R. Stimpson, E. Dixler, M. J. Nelson, & K. B. Yatrakis (Eds.), *Women and the American city* (pp. 185-211). Chicago: University of Chicago Press.

West, G. (1981). *The NWRO: The social protest of poor women.* New York: Praeger.

Wilson, E. (1991). *The sphinx in the city: Urban life, the control of disorder, and women.* Berkeley: University of California Press.

4   Women Organizing Their Place in Restructuring Economies

LOUISE JEZIERSKI

Economic restructuring within an emerging global economy has put women's roles in the local economy and community in flux. Since World War II, women have entered the paid labor force in increasing numbers, and this has had far-reaching consequences for the social division of labor. Women have responded by asserting new agendas and strategies for access to economic and political resources. This chapter reviews some of the major changes in the gendered local political economy and discusses their implications for future metropolitan development. The analysis[1] is based on both primary and secondary research on the transformation of two old industrial metropolitan economies to postindustrialism. The first case is Boston, where women's participation in traditional manufacturing has been the highest, historically. The second case is Pittsburgh, where women's labor force participation has been low historically because of the preponderance of durable goods manufacturing there but has recently increased because of the precipitous decline of this sector. In both cities, postindustrial economies are emerging based on growth in business, education, and medical services, industries that have opened some niches of employment for women.

However, the role of women in the postindustrial city is a precarious one because the gendering of jobs and urban spaces tends to peripheralize women and their families. The confluence of patriarchal work environments and the invisibility of women in the urban development process have worked together to marginalize women in the city. Nevertheless, women are central to the functioning of the new postindustrial

AUTHOR'S NOTE: I would like to thank Judith Garber, Robyne Turner, and Barry Goetz for editorial support in writing this article.

political economy. Important structural and political changes have affected women's roles in work, in the community, and at home. The gains women make will depend on their abilities to organize a place for themselves in the restructured social division of labor.

■ Women in the Post-Fordist City

Since the 1960s, both the Pittsburgh and Boston areas have experienced similar changes in their economies, in household formations, and in civic relations. The macrostructural forces that are taking effect in both regions and that are responsible for changing conditions in the realms of production, consumption, and social regulation, are described as a transition from Fordist industrialism to postindustrialism or post-Fordism (Harvey, 1987; Schoenberger, 1988).

The Fordist city was organized on mass production and consumption, although labor and housing markets were segmented by gender and by race (Bonacich, 1972; Gordon, Edwards, & Reich, 1982; Sabel, 1982). Patriarchy in conjunction with capitalism consolidated labor and households in mutually dependent networks of home and work that were organized as separate spheres of public and private life, spatially segregating men and women at work and at home (Hartmann, 1976). Most models of Fordist regimes of accumulation include a characterization of men engaged in core manufacturing, making a "family wage" that would allow women to take care of the household. This description held true for most middle-class families in the post-World War II period, although not so for working and poor families, where women have always worked outside the home. Women filled in on a part-time or temporary basis or in specific manufacturing and service niches, such as garment trades, food preparation, clerical work, or domestics (Milkman, 1991). Spatially, women found themselves isolated at home with their own children, or with others' children, in gendered suburban cul-de-sacs or urban enclaves, such as public housing units (Saegert, 1981). Wilson (1987) reports that 70% of the residents of the Robert Taylor homes on Chicago's south side are children. The subsidization of suburban infrastructure through tax breaks and federal expenditures for highways, cheap housing finance, and new schools fostered the isolation of the nuclear family (Checkoway, 1986). Thus industrial practices and federal policy worked together to segregate home from work life and worked to impose a gendered division of labor between women in the private sphere and men in the public sphere.

The problem of the invisibility of women has been exacerbated by the "placelessness" assigned to women. American urban researchers and urban planners often ignore the lives of families, women, and children as a distinct sphere of analysis. Cities are viewed as an unsuitable environment for children. The result of a good deal of "urban planning" sacrificed neighborhoods for downtown development—a process that worked to the detriment of women and children. Even the business functions of the downtown were biased to (male) commuters, with women's safety and comfort left unconsidered, even though women make up 98% of the clerical army.

A reconfigured city is emerging with post-Fordist arrangements of production and social reproduction. The postindustrial city is characterized by a loss of large factories and their workers, in a process described generally as *deindustrialization* (Bluestone & Harrison, 1982; Wilson, 1987). New regimes of flexible specialization (Piore & Sabel, 1984) or flexible accumulation (Aglieta, 1979) introduce a new economic order built on specialty manufacturing and personal and business services, especially in the sectors of retail, finance, insurance, real estate, and education. The conception of the post-Fordist city reflects fundamental changes in production techniques and the organization of the labor process too. Standardized mass production arrangements are replaced with small-batch production, automation, and craft work. Thus the "politics of production" are transformed. Hierarchical management styles and collective bargaining arrangements are replaced with quality control circles, specialized job learning, and greater management flexibility (Milkman, 1991). Consumption patterns are altered as well, from mass consumption to niche markets.

In the economic crises beginning in the mid-1970s, the Fordist social contract had to be renegotiated (Milkman, 1991). Women's work and the women's movement are fundamental to post-Fordist arrangements. Economic restructuring has brought the wholesale demise of key industries in the areas while at the same time new ones have grown. Durable-goods production, especially steel in the Pittsburgh area, and shoes and textiles in the New England region, have been lost to producers overseas. Corporations moved to substitute union labor with a "reserve army of labor" that was often dominated by females. Women workers in Mexico, the Caribbean, the Philippines, and Korea were installed in the "global assembly line." At the same time, many in the middle class, especially those who had jobs as managers in high-wage manufacturing sectors or who were victims of corporate consolidation and "downsizing," saw their jobs eliminated as well.

The deindustrialization thesis would predict that plant closures due to restructuring or recession in the 1980s would disproportionately affect women. Those who have long-held manufacturing jobs, such as in textile or electronics factories that have traditionally hired women, have found their well-paying, sometimes unionized, jobs disappearing because of plant closures or increasing subcontracting practices. Women who have recently acquired manufacturing jobs, such as in the steel industry, which opened up production jobs to women in the 1970s, found that they were the last hired and first to be laid off. Access to stable and well-paying manufacturing jobs has become increasingly difficult (Harrison, 1985; Harrison & Bluestone, 1988; Wilson, 1987).

A second, more optimistic thesis of postindustrial change, such as flexible specialization, would predict that women might benefit from new jobs connected to growing industries, such as those in high-technology and service industries, especially in the areas of health care and education. However, many of the new service jobs pay less than do manufacturing jobs. Moreover, restructuring has brought about increasing bifurcation in jobs and income between highly skilled and highly paid professionals and part-time, low-skilled, and underpaid service workers (Bradbury, 1985; Graham & Ross, 1989; Mollenkopf & Castells, 1991). A third point of view argues that structural inequality between men and women can derail any advantages from restructuring (McDowall, 1991). Tienda, Ortiz, and Smith (1987) found that despite industrial restructuring that can bring about new opportunities for women in occupations, gender gaps in earnings may be maintained. Moreover, changes in demand for women workers may be race specific: Affirmative action has brought increased opportunities for women of color, especially in the public sector (Epstein & Duncombe, 1991; Tienda et al., 1987), whereas immigrant women, often women of color, are reserved for part-time, seasonal assembly work (Glenn, 1985; Zabella, 1984).

Thus economic restructuring has brought about a fragmentation of experiences for working women. Although high-technology and professional specialty jobs flourish, work choices faced by many in the working class are limited to low-wage jobs and informal (often illegal) economies. Some women have lost their jobs and their communities in this restructuring process, whereas others have benefited from increased demand related to the professional and specialized skills that women offer, albeit often at lower wages. As a result, bipolar class formations emerge and create new schisms.

The post-Fordist city has been characterized as the "dual city" (Mollenkopf & Castells, 1991) or the "quartered" city (Marcuse, 1993). This describes the increasingly fragmented nature of urban life since World War II. Metropolitan areas are segregated into core cities and semiautonomous suburbs, whereas coherent and holistic community enclaves that once revolved around factory life no longer exist. New enclaves exist side by side, including underclass districts of hyperghettoization (Jaret, 1991), new immigrant neighborhoods, and gilded and gentrified blocks. With the reemergence of unregulated and informal economies in the urban political economy, corporate cities of the core are becoming "peripheralized" (Sassen, 1987). "Civitas," the political culture and social order of the local community under post-Fordism, is represented by increased insecurity, commodification, pastiche, and spectacle (Harvey, 1987). The local state has responded to the economic crisis by scaling back on welfare aid and low-cost housing for families, straining supports for education, and replacing public sector jobs while relying on privatization. The post-Fordist city is increasingly governed through entrepreneurship and public-private partnerships between the local state and the local business sector that leave much of the decision-making process in private hands (Jezierski, 1990).

The impacts of this transformation have been especially devastating for urban women. With the loss of manufacturing jobs, whole communities have been devastated, and ways of life have been deracinated. The family has borne a good deal of the heartache and uncertainty that economic restructuring has wrought (Hochschild, 1989; Stacey, 1991). Women at home went to work as men's wages could no longer promise financial security for the family. Since the onset of the crisis of Fordism and economic restructuring, women are choosing to join the labor force, even women with young children. The female labor force has changed from "working daughters to working mothers" (Lamphere, 1987; see also Berman, 1989). This has profound implications for the economy, the household, and the community.

Women entered the workforce at precisely the time that the social contract governing social welfare policy—housing, child care, schooling, and aid to children—was severely undermined. Thus a double burden was created, resting primarily on the shoulders of women who took on a "second shift" at home after the close of the working day (Hochschild, 1989). The second shift created struggles between men and women over household and child care, as well as worries over children who may spend time home alone after school. The spatial isolation of the suburbs further exacerbated the strains of bringing women into the metropolitan workforce. These issues are often ignored when researchers study only the restructuring urban political economy (McDowall, 1991).

It is crucial to understand how structural change has affected women and how fragmentation of the urban experience creates barriers and opportunities for women to actively respond to these conditions. All social environments are created through the interactive effects of both structure and social action, and the transforming city is a result of a process of "structuration" that coordinates contending interest groups, such as class fractions and gendered groups (Moos & Dear, 1986; Sayer & Walker, 1992). Given the conditions of macrostructural economic change, women are renegotiating the social contract that informs their lives in a restructuring American metropolis. In response to new challenges, women have demanded innovations in local institutions, such as day care facilities and an expanded role for schools. Under Fordism, women were not incorporated into the core of city politics as formal political candidates or in political party leadership. But within the peripheral political world, women were engaged in agencies such as nonprofits, in volunteer work, in behind-the-scenes political support, and grassroots organizing (Bookman & Morgen, 1988; Ferguson, 1989). They are also engaged in school and church community activities, and play an increasingly important role as consumers of "big-ticket" items, such as automobiles and major appliances. Women are transforming this power into having a greater say in urban politics, both at the grassroots level of urban social movements and in running for office. These "peripheral" cultural or social arenas are the new sites of struggle under post-Fordism and are increasingly becoming central political issues. Because women have largely worked in these intermediate structures already, their political influence may increase in the restructuring city.

Regional variations can illuminate these trends. In a comparison of Boston and Pittsburgh, I examine whether the historical role of women in the labor market and the public sphere makes a difference for their incorporation in the post-Fordist city.

■ The Boston Metropolitan Area

The story of women in New England reflects both traditional patterns of work and social life as well as new structural realities. Women and

TABLE 4.1 Women's Labor Force Participation Rates (in percentages) for
Selected Cities, 1890-1990

	1890	1900	1910	1920	1960	1980	1990
Boston	n.a.[a]	30.2	n.a.	36.9	42.7	52.7	60.6
Lowell, MA	37.2	41.9	46.0	42.9	41.5	51.8	n.a.
Pittsburgh	18.6	20.3	24.9	24.8	33.9	44.7	51.0
U.S. average	18.9	20.6	25.4	23.7	37.7	51.5	57.5

SOURCES: Data for Lowell, Massachusetts, Pittsburgh, and U.S. averages, 1890 to 1920 from
Greenwald (1989, p. 36, table 1). U.S. figures from U.S. Bureau of Labor Statistics (1991). Other data
from U.S. Bureau of the Census (1920, tables 87, 88; 1960, table P-3; 1980, table 120). 1990 statistics
are taken from the U.S. Bureau of Labor Statistics (1990, table 23), for metropolitan areas, percentage
of women in the civilian labor force. Note that these are not directly comparable across columns.
a. Not available.

immigrants have played an integral role in the New England economy
since colonial times. New England was the birthplace of the American
Industrial Revolution, and women worked in these first factories (as
early as 1790 in Slater Mill in Pawtucket, Rhode Island). Women's labor
was crucial to the organization of textile and shoe production, such as
that found in the paternalistic factory system of the mills of Lowell,
Massachusetts. Indeed, the New England region has long been characterized by the highest labor participation rate for women (see Table 4.1).
High rates of unionization were also characteristic of the region, although manufacturers began to abandon this region as early as the
1920s, moving textile production to southern nonunion shops (Harrison,
1985; Lamphere, 1987). However, women worked steadily as an integral part of the workforce, fulfilling a crucial niche because they were
paid less than men. Women fulfilled "flexible" employment needs as
temporary, seasonal, and part-time workers, and, often, they were paid
according to a piece rate system (Rosen, 1987).

The traditional manufacturing base of New England, especially textiles, shoes, shipbuilding, and metal work, had declined by the 1960s.
Thus core manufacturing jobs were lost when restructuring began by
the 1970s.[2] Post-Fordist changes in the New England economy reflect
similar changes in the national economy (Knight & Barff, 1987-1988)
with service jobs leading the way in job growth. However, New England
has maintained a higher share of jobs in manufacturing compared to the
nation as a whole. The new economy was built on growth in computer
manufacturing, research and development associated with area universities, and business services in the post-World War II years. Banking

and insurance were leading sectors in the 1960s, and by the 1980s, high-technology sectors in industries related to defense, computers, health, and education spurred new economic growth (Browne & Hekman, 1988, p. 182). Additive growth brought a new round of real estate speculation, construction, urban gentrification, and new retailing and personal services, such as restaurants. Some manufacturing was maintained, not because of geographical location or natural resources, but because of low costs of production, new technologies, and the area's advantages in skilled labor, especially its large percentage of workers with college- and graduate-level educations (Browne & Hekman, 1988, p. 177; Frankel & Howell, 1988; Harrison, 1985). (An astounding 20% of the labor force in the city of Boston has completed 16 years of education or more.) The defense-related industries employed many laborers skilled in electronics and computer technology, although in these industries, low-skill assembly work is also prominent.

In the 1960s, at the height of Fordism, Boston metropolitan area women made up 36% of the total labor force, and 38% of all women were in the labor force. Of working women, 42% were married with husbands present, and about 7% of women with small children worked outside the home. The largest occupational category for women was clerical work (36%, which is unsurprising, given the fact that the Katherine Gibbs School for secretarial training was founded in Providence, RI, in 1911). About 16% worked as operatives, reflecting the long-standing role of women in manufacturing in the region. Although manufacturing composed 29% of all jobs, textiles contributed to 10% of manufacturing jobs—a sector in which women traditionally worked. Fourteen percent worked in professional and technical occupations.

With post-Fordist restructuring, women found themselves at both ends of a polarized economic structure but maintaining their high labor force participation rate. By 1991, 65% of New England women were participants in the labor force, compared with 57% for all women nationally. As manufacturing opportunities have decreased for men and their wages have eroded, married women have joined the labor force to make up for losses to household income. Much of the decline in Boston's manufacturing jobs can be explained by the New England regional recession. In 1986, 10% of women here worked as factory workers. This number declined to 7.5% in 1991, signaling a historical reversal of New England women working in manufacturing. Whereas single women always worked in the factories of New England, their union wage jobs have also been lost. Women in New England are more likely to work in durable goods manufacturing than are their national counterparts. However, structural changes and recession have reduced the number of female operators, fabricators, and laborers: Although the national average for women working as operators, fabricators, and laborers is 8.1%, only 4% of Boston area women hold such jobs (U.S. Bureau of Labor Statistics, 1993). Nevertheless, there is still a considerable demand for female factory workers in labor-intensive industries, including new high-technology manufacturing and in traditional industries such as textiles (Browne & Hekman, 1988, p. 183). But these jobs are often part-time and seasonal and are relegated to immigrants (Ribadeneira, 1989; Walsh, 1983).

Female labor force participation in New England increased in the postindustrial sectors, especially in executive, professional, managerial, clerical, and retail sales occupations. Predictably, the largest occupational category for women in 1990 was clerical and administrative support (about 27%, equal to the national average). However, the second largest occupational category for women in the metropolitan area in 1990 was professional specialities—almost 22% of all working women (which exceeds the national average of 15.6%) and followed by another 16% in executive, administrative, and managerial occupations (the national average in 1990 was 11.4%). Clearly, women have been upwardly mobile since the 1960s. Much of the gain in occupational status has been located in service industries, reflecting the large number of colleges and universities and hospitals in the area. In figures available for 1980, about 43% of employed women in the region worked in services, especially medical and educational, compared to 39% nationally. Within this category, 44% of jobs were in hospitals or related health; 39% were in education; a significant 9% were in social services, religious, and membership organizations; and another 8% were in legal, engineering, and other professional services. (Comparable figures are not yet available for 1990.) Clearly, as women pursue educational advancement, they can find payoff in the labor force.

In civic life, women in the Boston area have been especially active in nonprofit-sector agencies. This sector can be considered part of the peripheral or informal economy, as well as the backstage of politics, but in the new entrepreneurial climate of the post-Fordist city, these areas are becoming more important. They also serve as fertile ground for preparing women for electoral office. Boston city politics are characterized as fractious and competitive and are still based on neighborhood ties (Hardy-Fanta, 1992; King, 1981; Travis, 1986).

Representational change began in the 1970s, as women engaged in mobilizations over school integration, which resulted in a number of women being elected to the Boston School Committee and to the city council. By the 1980s, although women have not yet been able to capture many of the available electoral seats, opportunities have opened up; four women hold city council seats in 1994. This female leadership was shaped in neighborhood politics. The Democratic party has not developed women in leadership roles at the local level, although they have made progress at the state level. Future prospects depend on a greater institutionalization of women in both business and political party leadership positions.

The "Massachusetts Miracle" brought prosperity to the New England region and offered new opportunities in postindustrial growth areas such as high technology, business and professional services, and education and health services. The growth economy in New England in the 1980s created an unemployment rate on average of about 4%, where almost every able-bodied person could work. The family poverty rate was cut from 24% to 9% in Massachusetts, whereas the national average of family poverty in cities had risen from 20% to 22% (Sege, 1989; see also Boston Foundation, 1989). Postindustrialism was viewed in the region as an antidote to deindustrialization. Although economic growth in the region should have promised women wider access to economic and social development, the impacts were both positive and negative for them.

■ The Pittsburgh Metropolitan Area

The historical images of the hardworking Three River City, the "Iron City", the smoky home of Pittsburgh steelers, and the headquarters of internationally powerful, industrial corporations such as Westinghouse, U.S. Steel, Alcoa, and Pittsburgh Plate Glass (PPG Industries), seem to overpower and overshadow the role of the women living there. Greenwald (1989) argues that historically, "Women in Pittsburgh had fewer options for wage earning and fewer opportunities for civic leadership" (p. 34). Women seemed to be excluded, both in the primary productive industries of the region and in politics, where a formidable public-private partnership was forged between a Democratic party machine and a Republican corporate managerial elite.

Until the turn of the century, employment for women was limited, and most working-class women were employed as domestics or worked in cotton mills. But although iron and steel mills employed about one third of all workers in 1910, these jobs were highly segregated by extreme class, gender, ethnic, and racial distinctions (Greenwald, 1989, p. 35). Women were discriminated against in these industries and composed only 10% of all workers in manufacturing. In 1920, only 14% of employed women in Pittsburgh worked in industrial manufacturing, compared to women in other industrial cities (Buffalo and Detroit at 22% and Cleveland at 25%), probably because the industrial base was less diversified in Pittsburgh. The discrimination against married women in the labor force left them few options, except for taking in laundry and boarders (Greenwald, 1989, p. 38). Domestic and personal services employed almost a third of all women in the workforce. However, by 1920, white-collar jobs provided new opportunities, and women worked as clerks in department stores, as clericals in offices (26% of all women workers), and as telephone operators.

During the 1960s, women still lagged behind in labor force participation, in contrast to both Boston and national averages. Here women made up about 29% of the total labor force. Only 28% of all adult women worked outside the home. About 13% of married women with husbands were working, and only about 2% with children under the age of 6 were employed. About two thirds of all working women were in some aspect of white-collar work: 15% of all working women were employed as professional, technical, or kindred workers; a third were clerical workers; and 11.5% were in sales. In a metropolitan economy where 37% of the total jobs were in manufacturing (half of these in metal industries), only 8% of women worked as operatives.

By 1990, 51% of women in the Pittsburgh area were in the labor force, compared with 60% in the Boston metropolitan area. Women in both areas were employed in similar sectors: In Pittsburgh, the largest occupational category for women is clerical and administrative support (29% for women in Pittsburgh). Again, the second largest category is professional specialties (17% in Pittsburgh); however, fewer women work as executives and managers, falling short at 10.6% whereas the national average is 11.4%. By 1990, both Boston and Pittsburgh had lost their lower-skill manufacturing jobs; in Pittsburgh only 5.3% of women were employed in this category, compared to 8.1% nationally (U.S. Bureau of Labor Statistics, 1992, table 25; U.S. Bureau of Labor Statistics, 1993).

The invisibility of women dissipated only with the demise of steel and the growth of new industries and a restructured local political regime. Women steelworkers were just getting a toehold in the exclusive

TABLE 4.2   Percentage Distribution of Employed Civilians by Occupation and Sex for Metropolitan Areas: Boston Primary Metropolitan Statistical Area and Pittsburgh Consolidated Metropolitan Statistical Area, 1991

	Boston Total	Men	Women	Pittsburgh Total	Men	Women
Executive, administrative, and managerial	18.2	20.4	15.8	13.0	15.0	10.6
Professional specialty	20.9	20.2	21.8	15.5	14.3	16.9
Technical and support	4.3	4.3	4.3	3.8	3.7	3.9
Sales	11.6	11.8	11.5	12.0	9.8	14.4
Administrative support and clerical	16.7	6.7	27.5	16.7	5.8	29.2
Service	12.2	10.9	13.5	14.3	11.2	18.0
Precision, craft	7.7	13.5	1.6	10.4	18.5	1.1
Machine operators	2.7	2.8	2.6	4.6	5.9	3.2
Transportation	2.3	4.9	.5	3.8	6.7	.5
Laborers	2.5	4.1	.9	4.8	7.6	1.6
Totals	100.00					

SOURCE: U.S. Bureau of Labor Statistics (1992). *Geographical Profile of Employment and Unemployment for 1991*, Table 25, pp. 119-121.

industry in the 1970s and 1980s, when steel in the United States began to face tough foreign competition that led to reorganization of the mills and the layoffs of tens of thousands in the area. The devastation of milltowns all along the riverbanks crushed steelworking families, so the women of these households went out to work, often for the first time. Families and communities organized to feed people, to pay their mortgages, and to counsel each other. They had to find new jobs and new roles to fill in a Pittsburgh that was being rebuilt. Glass and steel corporate cathedrals replaced smokestacks. Pleasure boats skimmed the rivers, bypassing barges. And women commuted to the Point, downtown, and to the hills of Oakland to provide support staff to the area hospitals, law firms, and universities.

Employment shifts as a result of restructuring and long-term trends of increasing diversity in the industrial base have benefited women. Men have lost high-paying manufacturing jobs, whereas women have gained employment and wage increases as the service sector has expanded (Jacobson, 1987) (see Table 4.2). Although women's labor force participation is still below the national average, the rate of increase in the last 20 years has been substantial, especially for married women.

The traditional structure of the local economy has limited opportunities for women to contribute to planning and running Pittsburgh's redevelopment. Women in the business sectors especially feel that very few women are included in civic organizations, such as the formidable Allegheny Conference for Community Development, a corporate CEO membership organization that has been instrumental in Pittsburgh's public-private partnership for the last 50 years. The tightly knit organization of Pittsburgh's private sector has tended to exclude women because they were not represented in the city's industrial structure. The Democratic party has also been an important organizer of city leadership, and since the neighborhood struggles that began in the 1960s, women have made some inroads. A number of women have been elected to city council, and Sophie Masloff, who got her start in neighborhood ward politics, was elected mayor in the late 1980s. Women in neighborhood organizations have also boosted their representation in new, citywide partnership structures that are key to city development planning since the late 1980s. Thus a two-tiered leadership structure has been created, with the private sector and local state fashioning economic development for the region in the long term; this sector is still run by the male corporate elite. Meanwhile, the institutionalization of neighborhood interests in elected and quasi-political bodies, such as the city council, community development corporations, and the Pittsburgh Partnership for Neighborhood Development, bodes well for women's issues. Women have had important leadership positions at the level of the mayoralty, the City Planning Commission, and the Department of City Planning, thus incorporating women in the day-to-day operation of local decision making.

In retrospect and in general, the lack of experience that women have had in the local labor force did not seem to hinder them from gaining jobs in the emerging postindustrial economy. Those who were educated or active in third-sector, nonprofit organizations moved into the mainstream of the economy by 1980. Both men and women in manufacturing lost ground, however, and the impact on families has been devastating. The changing household will bring Pittsburgh women up to the national norm of working mothers and wives. Yet household income will not reach levels achieved before recession and restructuring. This has hurt the economic development of the region as a whole. The new Pittsburgh promises both economic and political opportunities for women in the coming decade.

■ The Future of Women in the City

The precarious nature of the restructuring post-Fordist city will make for uncertain and insecure economic conditions for women and their families for the foreseeable future. The postindustrial economy is not replacing jobs lost to deindustrialization: New service and high-technology jobs, which are the growth sectors, may be made up of temporary or part-time work, pay less, or be subject to restructuring themselves. The deep recession in New England of the late 1980s was not expected and does not bode well for the salvatory characterization of these new jobs. However, women are finding a new place for themselves in service and technical jobs. The demise of the heavily male-biased, Fordist production formation in Pittsburgh has opened up new opportunities for women in the labor force, even though the manufacturing jobs lost to the men in their lives contributed to financial ruin of many families and undermined community.

As women enter the labor force in increasing numbers, the effects on family life and community are profound. Unfortunately, I have touched on this aspect too little here because of lack of space; however, some of these issues are included in other chapters in this volume. More research should be done that makes connections between urban political economies and the spheres of social reproduction. Here, however, I would like to argue that as women gain higher-status positions in the local economy, they will also be gaining visibility in local decision-making processes, especially if public-private partnership arrangements become more the norm. The Neighborhood Partnership in Pittsburgh has become an important body for shaping the city's redevelopment agenda. It does not have regional influence, but as women in city leadership roles become more prominent, they will be included in the larger bodies. Most important, the style of politics will change. However, this does not happen automatically. Although Boston-area women have long been employed, this positioning in the political economy does not translate into public power. Instead, women must also organize within the civic arena, especially party politics. Efforts such as Leadership Boston or Leadership Pittsburgh, which provide networking support for people in local civic and business organizations, will take some time to institutionalize women's place, but they are a start. The fragmentation of postindustrial city politics may limit these efforts and decentralize women's issues as simply another form of interest group pluralism, but the crises of urban restructuring will provide the conditions for rethinking urban planning and urban redevelopment (Jones, 1990, Young, 1990). Women will be involved in the discussion.

NOTES

1. This discussion is derived from an ongoing project on the restructuring of Pittsburgh, where I have interviewed over 60 informants (Jezierski, 1990) and a recent study on metropolitan restructuring and the formation of ethnicity in Boston, Hartford, and Providence. Research in both studies involves demographic and economic analysis using data from the U.S. Bureau of the Census and the U.S. Bureau of Labor Statistics, archival data analysis of internal reports and newsletters from local businesses, community associations, and government, and in-depth interviews with key informants on local political economy.

2. The New England region lost 252,000 manufacturing jobs between 1968 and 1975. However, 225,000 new manufacturing jobs were gained between 1975 and 1980, with an estimated 45% of the growth based on high technology. The textile industry began to decline in the 1920s but declined dramatically after World War II. Textile employment in the region dropped from 280,000 in 1947 to 170,000 in 1954 to 99,000 in 1964 (Frankel & Howell, 1988, pp. 296, 298).

REFERENCES

Aglieta, M. (1979). *A theory of capitalist regulation.* London: New Left Books.

Berman, R. (1989). The feminization of the U.S. workforce. *Monthly Review, 41,* 1-11.

Bluestone, B., & Harrison, D. (1982). *The deindustrialization of America.* New York: Basic Books.

Bonacich, E. (1972). A theory of ethnic antagonism: The split labor market. *American Sociological Review, 37,* 547-559.

Bookman, A., & Morgen, S. (Eds.). (1988). *Women and the politics of empowerment: Perspectives from the workplace and the community.* Philadelphia: Temple University Press.

Boston Foundation. (1989). *In the midst of plenty.* Boston: Boston Foundation.

Bradbury, K. L. (1986, September-October). Prospects for growth in New England: The labor force. *New England Economic Review,* pp. 50-60.

Browne, L., & Hekman, J. (1988). New England's economy in the 1980s. In D. Lampe (Ed.), *The Massachusetts miracle: High technology and revitalization* (pp. 169-187). Cambridge: MIT Press.

Checkoway, B. (1986). Large builders, federal housing programs, and post-war suburbanization. In R. Bratt, C. Hartmann, & A. Meyerson (Eds.), *Critical perspectives on housing* (pp. 119-138). Philadelphia: Temple University Press.

Epstein, C. F., & Duncombe, S. (1991). Women clerical workers. In J. Mollenkopf & M. Castells (Eds.), *The dual city* (pp. 177-205). New York: Russell Sage.

Ferguson, K. (1989). Grass roots, new routes. *Women's Review of Books, 16,* 13-14.

Frankel, L., & Howell, J. (1988). Economic revitalization and job creation in America's oldest industrialized region. In D. Lampe (Ed.), *The Massachusetts miracle: High technology and revitalization* (pp. 295-313). Cambridge: MIT Press.

Glenn, E. N. (1985). Racial ethnic women's labor: The intersection of race, gender and class oppression. *Review of Radical Political Economy, 17*(3), 86-108.

Gordon, D., Edwards, R., & Reich, M. (1982). *Segmented work, divided workers: The historical transformation of labor in the United States.* Cambridge, UK: Cambridge University Press.

Graham, J., & Ross, R.J.S. (1989). From manufacturing-based industrial policy to service-based employment policy? Industrial interests, class politics and the "Massachusetts miracle." *International Journal of Urban and Regional Research, 13*, 121-136.

Greenwald, M. (1989). Women and class in Pittsburgh. In S. Hays (Ed.), *City at the point: Essays on the social history of Pittsburgh* (pp. 33-68). Pittsburgh: University of Pittsburgh Press.

Hardy-Fanta, C. (1992). *Latina politics, Latino politics.* Philadelphia: Temple University Press.

Harrison, B. (1985). Increasing instability and inequality in the "revival" of the New England economy. In H. Richardson & J. Turek (Eds.), *Economic prospects for the Northeast* (pp. 123-150). Philadelphia: Temple University Press.

Harrison, B., & Bluestone, B. (1988). *The great U-turn: Corporate restructuring and the polarizing of America.* New York: Basic Books.

Hartmann, H. (1976). Capitalism, patriarchy, and job segregation by sex. *Signs, 3,* 137-170.

Harvey, D. (1987). Flexible accumulation through urbanization: Reflections on post-modernism in the American city. *Antipode, 19,* 260-286.

Hochschild, A. (1989). *The second shift: Working parents and the revolution at home.* New York: Viking.

Jacobson, L. (1987). Labor mobility and structural change in Pittsburgh. *Journal of the American Planning Association, 53,* 438-448.

Jaret, C. (1991). Recent structural change and U.S. ethnic minorities. *Journal of Urban Affairs, 13,* 315-330.

Jezierski, L. (1990). Neighborhoods and public-private partnerships in Pittsburgh. *Urban Affairs Quarterly, 26,* 217-249.

Jones, K. (1990). Citizenship in a woman-friendly polity. *Signs, 15,* 781-812.

King, M. (1981). *Chain of change: A struggle for Black community development.* Boston: South End Press.

Knight, P., & Barff, R. (1987-1988). Employment growth and the turnaround in the New England economy. *Northeast Journal of Business and Economics, 14,* 1-15.

Lamphere, L. (1987). *From working daughters to working mothers: Immigrant women in a New England industrial community.* Ithaca, NY: Cornell University Press.

Marcuse, P. (1993). What's so new about divided cities? *International Journal of Urban and Regional Research, 17,* 355-365.

McDowall, L. (1991). Restructuring production and reproduction: Some theoretical and empirical issues relating to gender, or women in Britain. In M. Gottdiener & C. Pickvance (Eds.), *Urban life in transition* (Urban Affairs Annual Review, Vol. 39, pp. 77-105). Newbury Park, CA: Sage.

Milkman, R. (1991). Labor and management in uncertain times: Renegotiating the social contract. In A. Wolfe (Ed.), *America at century's end* (pp. 131-151). Berkeley: University of California Press.

Mollenkopf, J., & Castells, M. (1991). *Dual city: Restructuring New York.* New York: Russell Sage.

Moos, A. I., & Dear, M. J. (1986). Structuration theory in urban analysis: 1. Theoretical exegesis. *Environment and Planning A, 18,* 231-252.

Piore, M., & Sabel, C. (1984). *The second industrial divide.* New York: Basic Books.

Ribadeneira, D. (1989, June 5). Boom bypassing Massachusetts Latinos. *The Boston Globe,* p. 1.

Rosen, E. I. (1987). *Bitter choices: Blue collar women in and out of work.* Chicago: University of Chicago Press.

Sabel, C. (1982). *Work and politics: The division of labor in industry.* Cambridge, UK: Cambridge University Press.

Saegert, S. (1981). Masculine cities and feminine suburbs: Polarized ideas, contradictory realities. In C. Stimpson, E. Dixler, M. J. Nelson, & K. B. Yatrakis (Eds.), *Women and the American city* (pp. 93-108). Chicago: University of Chicago Press.

Sassen S. (1987). Formal and informal associations: Dominicans and Colombians in New York. In E. M. Chaney & C. R. Sutton (Eds.), *Caribbean life in New York City: Sociocultural dimensions* (pp. 278-296). New York: Center for Migration Studies.

Sayer, A., & Walker, R. (1992). *The new social economy: Reworking the division of labor.* Oxford, UK: Blackwell.

Schoenberger, E. (1988). From Fordism to flexible accumulation. *Environment and Planning D: Society & Space, 6,* 245-262.

Sege, I. (1989, July 17). Study: Massachusetts poverty rate stalled. *The Boston Globe,* p. 1

Stacey, J. (1991). *Brave new families.* New York: Basic Books.

Tienda, M., Ortiz, V. & Smith, S. (1987). Industrial restructuring and earnings: A comparison of men and women. *American Sociological Review, 52,* 195-210.

Travis, T.M.C. (1986). Boston: The unfinished agenda. *PS: Political Science and Politics, 19,* 610-612.

U.S. Bureau of the Census. (1920). *Census of the population.* Washington, DC: U.S. Government Printing Office.

U.S. Bureau of the Census. (1960). *Census of the population.* Washington, DC: U.S. Government Printing Office.

U.S. Bureau of the Census. (1980). *Census of the population.* Washington, DC: U.S. Government Printing Office.

U.S. Bureau of Labor Statistics. (1990). *Geographical Profile of Employment and Unemployment.* Washington, DC: U.S. Government Printing Office.

U.S. Bureau of Labor Statistics. (1991). *Working women, a chartbook* (Bulletin No. 2385). Washington, DC: U.S. Government Printing Office.

U.S. Bureau of Labor Statistics. (1992). *Geographical Profile of Employment and Unemployment for 1991 (August 1992).* Washington, DC: U.S. Government Printing Office.

U.S. Bureau of Labor Statistics. (1993). *Working women in New England: 1991* (Publication No. USDL-024). Washington, DC: U.S. Government Printing Office.

Walsh, V. M. (1983). *Working women in Rhode Island: A study of female benchworkers in the costume jewelry industry.* Unpublished doctoral dissertation, Boston University.

Wilson, W. J. (1987). *The truly disadvantaged.* Chicago: University of Chicago Press.

Young, I. (1990). *Justice and the politics of difference.* Princeton, NJ: Princeton University Press.

Zabella, P. (1984). The impact of sunbelt industrialization and Chicanas. *Frontiers, 8,* 21-28.

5 Women and Collective Action in Urban Neighborhoods

GORDANA RABRENOVIC

Drug trade, street violence, inadequate city services—for many this is a picture of the typical American city. Cities often seem to be dangerous places not fit for living or for raising children. This is not a description of large cities only or solely a consequence of anti-urban biases. In the last 20 years, large and small cities have shared many problems, such as unemployment, poverty, drug trade, an increase in crime and homelessness, and the deterioration of local services. Current urban problems are due, in large part, to the loss of manufacturing jobs; the shrinking tax base, which has reduced the city's capacity to provide services to their residents; and the attraction and expansion of the suburbs for middle-class residents and businesses.

Nonetheless, many poor families with small children still live in our cities. Because of high rates of divorce and an increase in the number of single-parent households, women often find themselves solely responsible for both the economic and emotional well-being of their children. Trapped in their neighborhoods by poverty, they have taken action to stop violence against themselves and their children; ensure enough resources to feed, clothe, and house their children; and gain the right to be heard in the political arena. Here, I look at women's responses to neighborhood problems and at their ability to protect their neighborhoods from further deterioration, secure better local services, and make their neighborhoods safer.

AUTHOR'S NOTE: The research in Chelsea was supported by the Research and Scholarship Development Fund Grant (1992-1993) from Northeastern University at Boston. Lori Rosenberg, a graduate student at Northeastern University, assisted in conducting interviews and gathering the data.

This chapter examines women's organizing campaigns in two low-income neighborhoods.[1] In Schenectady, New York, many neighborhoods have deteriorated as manufacturing jobs were lost and spending for local services declined. Faced with an increase in illegal activities (drug trade, crack houses, fatal shootings) and the lack of police protection, women formed an organization to lobby for closing crack houses, obtaining faster police intervention, and imposing tougher sentences for drug-related crimes. In Schenectady, women showed that they can be active leaders and participants in local collective actions to improve neighborhood conditions, despite major obstacles in their communities.

In Chelsea, Massachusetts, women's participation in local struggles has been prompted by their sense that their families and culture are in danger from changes in the delivery of public services. When Boston University took over the public school system and proposed a mandatory preschool and extended day care program for preschool children, Latina women and Latino men mobilized to oppose it because they saw it as a threat—Boston University was trying to supplant their families and their culture. As a result of this anti-Boston University mobilization, a Latina woman was elected to the school board for the first time. Today Latina women in Chelsea actively participate in local political decision making as members of the Bilingual Education Parents Advisory Committee and the Commission on Hispanic Affairs.

This chapter offers further evidence that issues facing our cities are seen by women as crucial for the survival and well-being of their families. Political participation by women in urban communities shows that they seek an active role in local decision making and service provision. It is important to note that by framing their concerns in terms of the needs and rights of women, men, and children, women are able to make a connection between their gender-based demands and their family and community issues. Consequently, the entire community becomes the constituency of women-led organizations and movements. That can make them a powerful force for social, political, and economic change.

■ Women and Place-Based Collective Actions

Residents in any geographical area share a number of interests. They enjoy or suffer the same quality of municipal services and use the same public facilities. They are affected by each other's actions and those of public authorities, and they are interested in ways of improving their neighborhoods. By pooling their resources and sharing costs through neighborhood groups, they seek to solve common problems and secure collective benefits (Crenson, 1983). The need for place-based collective actions has increased, argue Ley and Mercer (1980), because a large number of the new conflicts and popular demands are seen as place related, stressing the importance of territory and services. In this new political situation, place-based collective actions are sources of cross-class alliances, according to Castells (1983) and others in new social movement literature. Common consumption problems and issues potentially link different social classes together and are a source of their unity. The main interest that the members of local organizations share is their place of residence. This is a unifying force that, despite all differences, keeps them together.

Recognizing the importance of the place where people live and its contribution to the social well-being of its residents as a group has a long tradition in sociology, reaching back to the seminal work done by the Chicago School of Sociology (Park, Burgess, & McKenzie, 1925). People develop social ties and feelings about their neighborhood. Positive feelings influence mental and physical well-being, make residents less fearful of crime, decrease their residential mobility, and increase their local political involvement (Fischer, 1982). Some theorists have argued that emotional attachment is less important than in the past, in part because of residential mobility and the ability to satisfy daily needs in larger areas (Greer, 1962). However, other research points to the fact that, based on their community's opportunities, people will draw on their emotional and social resources to build lives and entrepreneurial schemes (Logan & Molotch, 1987). Emotional-social ties, such as chatting with neighbors and having a high proportion of friends and relatives in the area, are important sources of social support in the neighborhood (Wellman, 1979).

For people living in poverty, support networks and sharing resources are means of economic survival (Susser, 1988). In that respect, some neighborhoods provide an opportunity to resist economic and political marginality. Also, often, neighbors are informal agents of social control, are influential in the socialization of one's children, and help in promoting neighborhood safety. This is especially important for women because they are more likely than men to face social isolation. The reasons for that isolation may be their responsibilities in caring for young children, their lack of knowledge of the English language, or poor public transportation that makes it difficult to visit family members and friends who live outside the neighborhood. So the neighborhood is an important place for satisfying social and emotional needs and for providing economic assistance and opportunities for political mobilization.

Research on women's participation in political life documents that as early as the beginning of the 19th century, women were active members of place-based local voluntary associations. The goal of these associations was to increase the public provision of social services. Their association with home and family resulted in women's facing fewer cultural barriers and role conflicts if they were active in community politics. Women organized around their families' daily necessities, because caring for children and family was seen as a traditionally legitimate concern for women and because women had little opportunity to participate in other areas of public life (Wolfe & Stracham, 1988). In that respect, their concerns were primarily focused on the domestic, family, and environmental aspects of urban life. When women participated in movements that addressed political and economic concerns, they were more likely to campaign within their own communities or to support actions taken by men. However, evidence exists that women frequently initiated collective action, organized support, and held leadership roles. But, often, their success and visibility and the recognition of their struggles brought in men who took over the leadership (Lawson & Barton, 1990).

The new research on women's participation in local struggles challenges the assumptions of women's "traditional concerns." It shows that women face issues that reflect the interconnection of their family, work, and community roles and responsibilities. Women have organized around needs to improve education, health care, neighborhood safety, and social services (Morgen, 1988). Ackelsberg (1988) argues that by addressing these issues and putting them on the political agenda, women are legitimizing their demands and overcoming the "artificial" separation between public and private spheres.

Women in low-income neighborhoods also react to the political and economic changes that created instability in employment opportunities and increased residents' dependence on the state in providing life-maintaining services. When the state, as often happens, fails to respond or adequately provide needed resources and services, women mobilize communities to fight for those resources and services. In doing so, women become involved in formulating political demands and organizing for them. "They took leadership roles. They became presidents of local block associations; they made speeches, coordinated demonstrations, organized food distribution, and dealt with politicians—both local and national" (Susser, 1988, p. 267). This is also what happened in communities I studied.

■ Clean Sweep United

The major problem facing women and their children in Schenectady is poverty. Poor families suffer from the lack of adequate income, education, and decent housing. As the literature documents, living in poverty is stressful for families. Insufficient money and food and daily life in deteriorated and overcrowded housing directly influence the physical, psychological, and social well-being of women, men, and children (Wilson, 1987).

In Schenectady, individual poverty and deteriorating neighborhoods are related to the restructuring of the American economy. In the 1970s, General Electric, the major employer in the city, reduced its operation drastically. The changes in Schenectady's economic base lowered the city's tax base and led to a decline in job opportunities, the quality of local services, and real estate values, forcing many residents to move. In the past, residents survived tough economic times because they relied on ethnic solidarity and took care of each other. Both residential and occupational mobility have weakened ethnic bonds and their support functions. At the same time, competition for jobs and resources increased the conflict between ethnic or residential communities that still survived and left little room for them to organize to address neighborhood problems. Hamilton Hill, one of the poorest neighborhoods in Schenectady, is a good example of how economic restructuring inhibits collective action and contributes to the fragmentation of the community by decreasing individual resources and increasing neighborhood deterioration.

This neighborhood was always poor, but proximity to General Electric plants made it a convenient place for employees to live. When plants were closed, however, the residential exodus began. Residents who could afford to left for the growing suburbs, attracted by better services and larger houses. Older residents, whose children left for colleges and jobs in other communities, remained. Gradually, those displaced from more expensive cities moved into the neighborhood. The new residents—Puerto Ricans, Dominicans, and then Vietnamese, Cambodians, and Afghans—were very different from the previous ethnic groups.

They were not attracted by job opportunities but, rather, by low rents because they could not afford to live anywhere else. For the same reasons, the number of African American residents also grew in these neighborhoods.

The large concentration of a low-income population and declining property values brought new problems: drug-related crimes, prostitution, and steadily worsening local services. Also, a large number of buildings were owned by absentee landlords who did not maintain their housing. The conditions of the neighborhood and the economic prospects of the residents seem to correspond. However, Massey and Denton (1988) argue that the concentration of poor people in certain neighborhoods is not solely a consequence of economic factors. According to them, residential segregation is the primary reason that minority groups are concentrated in the most deteriorated neighborhoods in the city. Their social and economic well-being is then affected by the neighborhood's exclusion from city amenities, opportunities, and resources.

Hamilton Hill, like other poor neighborhoods, needs voluntary and social service organizations to overcome neighborhood problems, to provide support and a "sense of community." Unfortunately, many existing social service agencies have become weak or ineffectual because of a lack of finances, poor coordination, and a limited crisis management orientation. For a large number of women at the Hill, fighting for the improvement of services became a paramount task.

Many families depend on social service organizations to provide emotional, informational, and instrumental assistance. Researchers have found that, for example, teenage mothers who have access to low-cost child care services, family financial assistance, and vocational training can achieve economic independence (Mayfield, 1991). Similarly, welfare payments help women leave abusive relationships and poorly paid jobs (Naples, 1991).

Residents of Hamilton Hill lacked the individual and collective resources to improve the conditions of their lives. That forced them to find alternative ways to address their problems. The literature on community organizing and new social movements stresses the importance of sponsoring community organizations and using grassroots tactics and protest as a way to gain more power locally. Community organizations facilitate contacts between people in the neighborhood. Through mobilization and confrontation, participants gain a sense of togetherness that helps to establish unity and cohesion in the organizations. Because women who are leading organizations have experience with being powerless, they learn to use protest as a strategy to overcome powerlessness and address specific issues in their neighborhoods. Protest strategies can allow women a creative use of their resources linked to their traditional gender role as mothers. West and Blumberg (1990) describe how women mounted protests that gained attention and sympathy by using social stereotypes to challenge their opponents and symbols of their roles as wives and mothers, such as pots and pans, to publicize their demands and ridicule their opponents. With an imaginative use of familiar symbols and theatrical devices, women often were able to increase public visibility for their demands and win some state intervention.

This is the kind of strategy that women used in Schenectady. They formed Clean Sweep United, a citywide organization, to deal with the city's overwhelming problems with drugs. This organization had about 100 members, most of them White women living on Hamilton Hill. They failed to attract women from other racial and ethnic groups because many White women believed that the new residents brought crime to the neighborhood. Minority women and their children were also victims of crime and shared the same concerns for safety as did White women, but race and ethnicity had a stronger impact than did common problems.

Some leaders of Clean Sweep United were aware that they lacked the support of minority women from the neighborhood and made an effort to attract a more racially balanced membership. One of the members involved in outreach actions explained the reasons:

> We didn't want to be seen as an all-White group. In the beginning we had only one member who was Black. When you deal with drugs, it also becomes a race issue. On the Hill, everybody was sensitive to the drug problem, White and Black alike. Therefore, we started encouraging more Blacks and Hispanics to join us.

But that was not an easy task, and the organization continued to have primarily White members. The lack of cross-racial and cross-ethnic coalitions shows that for poor residents, race and ethnicity are powerful forces that often divide them and turn them against each other. Instead of uniting residents, the place becomes a contested terrain. By concentrating solely on consequences of poverty—drugs and crime—and not on its causes—unemployment and racial discrimination—White women ignored issues that were important for minority women and their families.

The primary focus of Clean Sweep United was to make Hamilton Hill a safer place to live. The members were frustrated by the lenient criminal justice system in Schenectady. In 1988, when the organization was formed, there had been 12 violent shootings in the Hill area. Drugs were dealt openly in the neighborhood, but not many arrests followed, and the police complained that the plea-bargaining system made their arrests fruitless. Members of Clean Sweep United felt that the city was trying to sweep the drug problem under the rug, and they wanted to make it difficult for the city to be content with "business as usual." Members, however, did not want to fight drug problems alone. They wanted to force the police to do the job.

By its members' blunt talk and actions, the organization was successful in getting media attention. They also made certain that local newspapers and television stations were informed about their actions. The purpose of the organization, according to one of its founding members and leaders, was to force the city and the district attorney's office to prosecute drug crimes more seriously:

> We tried to talk with the mayor first, but when she did not respond, a bunch of us decided to do something. We went to the mayor's house and sat on her stoop. "If we can't have lunch on our stoops we're going to have it on yours," we told her. Next, we went to the Schenectady County District Attorney's neighborhood and told his neighbors that we were going to continue to come back and disturb them until "their neighbor" stopped letting drug dealers loose [through the plea-bargaining system]. Well, the district attorney said that we didn't bother him, but his wife and neighbors were concerned. Nobody likes strangers wandering in their neighborhood.

These tactics worked. In the upcoming election, the district attorney lost, and the new district attorney stopped the practice of plea bargaining. As a consequence of this change, about 100 more arrests were made. Schenectady police arrested so many people that their jails could not hold them, and they had to rent trailers. The organization's members were especially pleased that the two biggest drug dealers were sentenced to 15 years to life in jail.

The organization's actions were not applauded by everyone. Some public officials and residents accused the organization of being radical and, because of women in leadership positions, hysterical. Fighting for police protection is hardly a radical goal, but it was unusual because women were active and held leadership roles. They accomplished the major goal of the organization: to make the city and its residents aware of the drug problem. The members believed something had to be done and felt that they were taking risks to save their neighborhoods and their families. One woman, a single mother and one of the organization's more vocal members, described her involvement:

> Participating in this organization was dangerous. We would go and beam a light on the house where drugs were being dealt to prevent customers from coming in. Or we would take license numbers and give them to the police. Sure, I was scared, but I was more scared by the prospect of my son being shot playing in the backyard. That's why I decided to do something. Better to be shot doing something than doing nothing at all.

Some civil rights activists from the neighborhood also did not agree with the organization's actions. They believed that the Clean Sweep focus was too narrow. One of the activists who worked in Hamilton Hill for years said,

> Drugs are manifestations of other problems, such as poverty, powerlessness, and unemployment. More sweeping law enforcement, more prisons, and stiff sentencing, although important, could not solve the problem. Instead, Schenectady needs more economic opportunity, especially for minorities, and more drug rehabilitation programs.

This opinion is shared by some researchers who studied local actions. Saunders (1981) has argued that protests over services tend to be transitory and fragmented. Also, organizing around specific locally based issues may prevent participants from seeing connections between their individual problems, the problems of their neighborhoods, and social conditions (Gottlieb, 1992) and, consequently, from linking their struggle to more enduring political movements. The real challenge to single-issue organizations is how to broaden their agendas as well as their membership base. Otherwise, they disappear from the local political scene.

This is what happened with Clean Sweep United. After the organization succeeded in achieving its main goal (tougher sentencing and no plea bargaining), it kept a low profile. However, not all drug-related problems were solved. Therefore, some members decided to continue their efforts by joining the neighborhood association, where they believed that they could continue to work on important neighborhood problems, especially those due to drugs. The neighborhood association was more racially balanced. The president and several members of the board were Black. The majority of the members of the neighborhood association were homeowners with a common interest in preserving their investment in the neighborhood, which allowed them to overcome racial and ethnic differences.

The consequences of these local struggles for the women that participated in Clean Sweep United went beyond making their neighborhood a safer place to live. These local struggles also had a positive effect on their sense of attachment to their neighborhood, their sense of community, and the development of their political consciousness. Through their collective efforts, these women met other people from their neighborhoods and began to feel that they belonged to a community, which they, in fact, were creating. As other researchers pointed out, political participation offers women the opportunity to cement friendships and to act rather than feel stifled by their frustration (Bookman & Morgen, 1988). The women who continued to be active in the neighborhood association began to think in more political terms. They began to see how their problems were related to the political and economic conditions of their communities, and they became more committed to changing these conditions.

■ Struggle for Education

In Schenectady, women organized because they wanted better services for their community. In Chelsea, Latina women went one step further. Their strategy was to gain control over local services. They were disturbed by the changes planned for their school system—Boston University was to take over the Chelsea public school system—and thought that those plans would affect their children negatively. The goal of their struggle was to force the schools to address the needs of all children in Chelsea and to become accountable to the parents. This strategy is similar to one used in the 1960s by working-class women from diverse ethnic and racial backgrounds who built powerful welfare rights and tenants movements to challenge their limited access to economic resources and political power (Pope, 1990). Although it had limited success, this strategy enabled women to gain political power to influence welfare policies and to participate in articulating solutions to housing problems.

In many respects, Chelsea is like Schenectady. Although close to Boston, it is a city with a declining economic base, widespread poverty, a high rate of illiteracy, and high dropout rates from school. It is an ethnically and racially diverse city and includes recent immigrants and migrants from various countries, such as Puerto Rico, El Salvador, Honduras, Guatemala, Costa Rica, Colombia, the Dominican Republic, Vietnam, and Cambodia, including a sizable undocumented population. Chelsea is highly dependent on the state for revenues and had such serious financial problems that in 1992 it went into state receivership.

Given such social and economic conditions, some city leaders turned to Boston University, believing that it could better manage the Chelsea schools. Boston University's School of Management was given a decade (1988-1998) to develop and implement a model of comprehensive urban education. Their proposal, titled "A Model for Excellence in Urban Education," recommended changes in the school curriculum, administration, and school structure, as well as in the provision of health care and social services. The university wrote that it would "assume authority and responsibility to assure that Chelsea's public schools become a national model of urban education which provides every student with the necessary knowledge, skills and values to make each a productive citizen" (Boston University, 1988, p. I-2).

An important part of the plan for educational reform was an emphasis on early childhood education. University officials believed that all students could enter kindergarten with the same skills and knowledge if they attended preschool classes for children 3 and 4 years old. They argued that children face problems growing up in "dysfunctional" families and that the school system has to provide them with extended school days. The plan also called for strengthening schools' ties to the community and for paying attention to the city's cultural diversity. Chelsea's "family school" concept of education meant that "each school is linked to a social service referral network that notifies appropriate community agencies of family problems, and keeps school personnel constantly informed on any actions taken" (Boston University, 1988, p. II-3).

From the beginning, however, the university and the community had differences with each other. It became clear to the community that the Boston University Management Team would have the final say in all decisions about education in Chelsea. The teachers were disturbed that the management team rather than the elected school board would have the power to hire and fire teachers. Also, Boston University, as a private institution, was granted exemptions from state laws requiring open meetings and public records and from any lawsuits brought against the school district. The teacher who was active in the teachers' union explained:

Here in Chelsea, the university got a sweetheart deal. All the power and no accountability. They said that as a private organization they have businesses that they have to protect, and this is why they do not want the records public. But for us it is important that we have access to our personal records and to financial records. This is a violation of the state constitution. You cannot mingle public and private funds. And I don't know why the city signed to be responsible for all legal costs if they [Boston University] did something wrong. . . . This reform is not about education; it is about governance.

The teachers were not the only ones critical of the university's plans for Chelsea schools. Despite a formal commitment to a multicultural approach to education, the university did very little to include representatives of Chelsea's ethnic communities in the negotiations. On a number of occasions, university officials even talked about saving money by teaching children English early, reducing the need for bilingual education. Those comments mobilized the mothers of Latino children. They felt that the university was challenging their ability to take care of their children. Some mothers were concerned that they would lose their children to schools that would make them learn English and forget Spanish. Others simply objected to being forced to send their "babies" to school all day.

From the beginning, the Latino community did not participate in the negotiations between Boston University and Chelsea because of their low status in Chelsea. Socially and economically, the Latino population as a group fares poorly in comparison with other residents of Chelsea. They are poorer, more likely to be employed in declining sectors of the economy, have higher rates of unemployment, and have lower levels of formal education and fluency in English (Kennedy & Stone, 1990). They are also more transient and live in more socially fragmented neighborhoods. Resource mobilization literature stresses that, to be successful, local movements need financial and political resources, such as money; access to information, knowledge, expertise; and sociopolitical support. Studies have shown that local groups use these resources to try to influence decision making (Crenson, 1983). It is hard for women who live in poor neighborhoods to mobilize needed resources. Because of a large concentration in a small number of low-paid, dead-end occupations, working-class men and women lack both economic and social resources, which undermines the power of their organizations.

However, Latina women in Chelsea were able to build on their personal relationships and were successful in using them for political purposes. They used social resources—support from families, kin, friends, and neighbors. In theory, *social resources* are defined as ties and connections between people both within and outside the neighborhood (Lin, Simeone, Ensel, & Kuo, 1979). Social ties are enhanced by the existence of common values and common needs. Another factor that contributes to developing social ties is the amount of individual cultural capital. *Cultural capital* is defined as the acceptance of values and norms of the larger society. Examples include organized democratic participation, knowing how institutions operate, and being able to use them to one's own advantage (DiMaggio & Mohr, 1985). The immigrants who have such capital can more easily become integrated into the American society and gain access to resources for themselves or their communities.

Hardy-Fanta's (1993) research on Latina women's participation shows that, often, immigrants successfully use their ethnically based personal connections, collective organizations, and political consciousness to mobilize their communities. They participate politically to make "connections between people, connections between private troubles and public issues, and connections that lead to political awareness and political action" (p. 3). It is, then, possible to expand the notion of cultural capital to include resources based on ethnic values and ties. Women use these ties not to create an ethnic enclave but, rather, to make a connection between their ethnic community and the larger society.

For a minority or immigrant community that suffers from a lack of political power and is most vulnerable to economic changes, these resources can have a strong influence on the success of locally based collective action. Similarly, studies of women's activities in the workplace have found that women's participation in friendship and ethnic networks and their sharing common churches, schools, and neighborhoods are instrumental in fighting for their work-based concerns (Sacks, 1988; Zavella, 1987). It is important to include these types of women's networks as resources because, as Ackelsberg (1988) suggests, women do not necessarily enter the local struggles as isolated individuals, "but as women strongly rooted in their class, ethnic, or cultural communities" (p. 303).

In Chelsea, Latina women had to overcome many obstacles to create a common organization. The Latino community is a very diverse community consisting of people from a number of different countries and socioeconomic conditions. Some common concerns about the status of Latinos in Chelsea helped to create cross-class and cross-national alliances. In general, Latinos had not been political participants; they did not even have a representative on the school board, although 54% of the students were Latinos. The public meetings held to discuss education reform were conducted in English and were not translated into Spanish.

Members of the Latino community shared similar concerns for bilingual education and wanted to have at least one Latino representative on the school board. That became the basis for their collective action. The mobilization of the Latino community was also facilitated by the support it received from some other organizations in the city. The teachers' union, seeking support in its struggle against Boston University, made Latino parents aware that bilingual education was in jeopardy. At the same time, the new leaders of the Commission of Hispanic Affairs, a Chelsea-based, nonprofit advocacy group, renewed their organization over this issue and decided that its priority was to fight for a greater say in community affairs. The commission found an ally in the Organization for Multicultural Educational Training and Advocacy, a Boston-area organization that suggested that the Latino community might gain more political power by participating in electoral politics in Chelsea. Mothers were particularly active in a voter registration campaign and helped elect the first Latina woman on the school board. They also worked on strengthening the work of the commission and were instrumental in forming a new Chelsea organization, the Bilingual Education Parents Advisory Committee (PAC). These women were aware that their fight was not just to win their children's rights but also to be taken seriously by school and city officials.

Until vocal members of these Chelsea-based organizations challenged Boston University's plans, it had been easy for the university and the Chelsea government to exclude the Latino community from any discussions about educational reform. Today, the Commission of Hispanic Affairs and the Bilingual Education PAC are making the Latino community interests heard in the city government, and Latinos are recognized as a force in the city's politics. The Latino community was successful in some of its efforts to change Boston University's plans for their children's schooling. But the problems with bilingual education continue because schools do not have enough bilingual teachers, class materials, or staff who speak Spanish. Parents are still intimidated by the school administration. A mother who is actively involved in the PAC described how difficult it is to negotiate with the school administration:

> Boston University's people and superintendent do not have respect for parents of children that they are supposed to teach. They don't want to

listen to us. It is hard for parents who don't speak English to come to the meetings without translation or to talk with teachers who don't speak Spanish.

The lack of resources in the city is creating a conflict between Latino parents and non-Latino parents of school children who do not support expenditures for bilingual education. The conflict adds to the social fragmentation of Chelsea and makes the struggle for a decent education much more difficult. This is similar to other studies that show how limited urban services pit racial and ethnic groups against each other (Luttrell, 1988; McCourt, 1977).

Some of the most recent complaints against Boston University came from all the parents, who objected to the university's strong emphasis on early childhood education. Parents and teachers of older children felt that these programs are given more financial support than the others. The class size is kept low, classes have teachers' aides, and there are enough supplies and classroom materials. However, even these classes do not have bilingual teachers, only bilingual teacher aides. One of the active educators in Chelsea said,

> They [Boston University spokespeople] say that early childhood educa-
> tion will prepare children for further education and that by intervening
> early we can solve the problems in the higher grades. But in Chelsea,
> people come and go, and the children in higher grades will not necessarily
> be the kids that have benefited from early childhood education. What
> would they do then? Besides, parents know that after the third grade the
> program is not good and that if they have economic means they will
> transfer their children to private schools.

The common criticism that older students are treated unfairly in the school system is bringing Latino and non-Latino parents together. Discrimination against older children cuts across ethnic and class lines, and with time, it can bring some unity among Chelsea parents. However, the conflict over bilingual education is also a struggle over the status of immigrants in Chelsea. In their struggle for their children's education, Latina women who participate in the PAC and the Commission of Hispanic Affairs are becoming aware that they are struggling to build Chelsea into a diverse and tolerant community. Historically, this community was a hub for immigrants who were expected to learn English and accept American values, attitudes, and norms. The role of education was to socialize newcomers with American ways. Today, the view that immigrants have to be assimilated within American society has been challenged by the advocates for a culturally pluralistic society. In such a society, the role of education is to nourish diversity and create opportunities for children to develop their potential.

■ Conclusion

These two cases provide evidence that a large number of women who live in poor communities initiate and actively participate in urban struggles. Alone or in a coalition with other groups, they react to the problems facing their residential communities. In the first example, we saw that changes in Schenectady's economic conditions have had a profound effect on life in residential communities, turning Hamilton Hill from a stable community into a transient and fragmented community. The major concern of residents was to stop further deterioration. This is a difficult task, because in declining industrial cities the resources for neighborhood improvement are limited. Impoverished neighborhoods are forced to compete among themselves for resources. Hamilton Hill had little chance for improvement. It faced a "double" disadvantage, being a poor neighborhood in a poor city. However, despite these obstacles, mostly White women responded to the neighborhood problems by using collective action. They were able to use their personal relationships to mobilize other women in the neighborhood. The use of protest strategy allowed them to gain public recognition and to force public officials to address their concerns: to protect their neighborhood and, consequently, their children and families. However, they were not able to overcome racial barriers. This is one of the most difficult challenges that urban movements in poor neighborhoods face. To overcome social fragmentation of the neighborhood, the White women of Hamilton Hill will have to develop new strategies to involve minority women in their activities.

That may happen as White women become more aware of the true causes of their neighborhood problems. The struggle with city officials helped some women to develop their political consciousness, to see beyond their own immediate concerns for the safety of their families and to build alliances with minority women. Their membership in the neighborhood association offered them the opportunity to work for the well-being of the whole neighborhood. They brought with them the skills and experience that made them valuable and important members of these new organizations.

In the second case, Chelsea, women did not have to learn that they needed to build an alliance but that they were part of the alliance from the beginning. They were successful in drawing on their social resources—ethnic ties and culture—to organize their ethnic community and to protect their family-based concerns. Over time, their goals widened: They no longer wanted simply to gain some control over services; they wanted to overcome the problems of political marginalization of the Latino population. Their strategy was to seek electoral representation and formal participation in the political process. Chelsea is a city where the feeling of powerlessness is overwhelming. Its schools have been taken over by a private organization not sensitive to parents' and community concerns. In light of these events, the mobilization of Latina women and their community is even more significant.

In Chelsea, Latina women hold leadership positions. They are the presidents of local organizations and members of executive boards. They articulate demands and participate in the search for solutions. They go to public meetings and speak at them. For them, issues of family and of ethnicity are interconnected. By fighting for a community tolerant of ethnic diversity they are also fighting for a community sensitive to family concerns. They also feel empowered by these new roles and responsibilities.

These two neighborhoods differ by how women perceive neighborhood problems and what they are willing to do about it. In Hamilton Hill, the focus of the conflict was the neighborhood, and collective action was based on a single issue: fight drugs to stem neighborhood deterioration. The strategy was to fight the neighborhood problem—crime—by pressing the local power structure to do something about it. Women sought unity based on their sharing the same neighborhood. In Chelsea, collective action was based on the struggle for school policies that respected Latinos' common language, their ethnic identity, and their family needs. Women were part of a much broader coalition defined by their ethnicity. The strategy was for Latinos to become part of the power structure. In this case, ethnicity was an important force in motivating local collective action. Women also used their rights as parents to be heard in decision making that affected their children.

For women in both communities, participation in collective action was empowering. It allowed them to play a more active role in public life, increased their skills, and turned them from victims to active participants looking for solutions to neighborhood problems. These examples show that women can be successful in legitimizing their concerns and putting them on political agendas. These women were able to recognize the forces that created their problems and were working toward changing them. Also, they were able to compensate for their lack of resources—money, prestige, formal education, well-paid jobs—by using informal networks, grassroots mobilization, and coalition building.

However, there is a limit to how much women can obtain for their communities using these techniques. Clearly, to sustain any gains that they made, women have to work toward improving the social and economic conditions of their communities. Rebuilding the neighborhood has to be connected with larger redevelopment policies, such as the creation of meaningful, skill-based jobs for the residents, and the improvement of urban infrastructure, housing, education, and health care. In Latin America, for example, this is already happening. As women organize around consumer and service demands, the need to gain jobs and housing for themselves is changing "from individual survival strategies to collective ones that place pressure on political parties and government bureaucrats" (Stephen, 1992, p. 90). Their collective action, in part, depends on political opportunities made available by states, as well as on changes in identity as women participate in grassroots mobilization. The goal of women participating in a community struggle, then, becomes to challenge the social organization of society (Lind, 1992).

Women still have, and need to address, other issues that affect their well-being—the questions of equal opportunity and economic advancement for women, the prevention of sexual harassment at work, equal pay for equal work, and the protection of sexual and reproductive rights. In this respect, feminist organizations, which are more engaged in fighting for solutions to these problems, can be of help to women participating in a community struggle. By addressing issues of mutual interests, these two groups of women can also share resources. Gender-based organizations often have more economic resources, organizational skills, and technical knowledge than do community-based organizations. Women in gender-based organizations are also better connected and have greater access to information. By improving their positions at work, women can more successfully address their community-based concerns.

This means, as Ackelsberg (1988) argues, that women have to address simultaneously their multiple roles and responsibilities across ethnic, racial, class, and gender realms. Women in Schenectady and Chelsea started to do that. They saw their constituency as the entire community. If they move from issue-specific struggles to finding the causes of urban problems, their organizations could become forces for social, political, and economic change.

NOTE

1. Data for these two case studies were collected through survey questionnaires and personal interviews. I also used secondary sources: local newspapers and official documents. The first case study was completed in 1990, and the second is an ongoing project.

REFERENCES

Ackelsberg, A. M. (1988). Communities, resistance, and women's activism: Some implications for a democratic polity. In A. Bookman & S. Morgen (Eds.), *Women and the politics of empowerment* (pp. 297-313). Philadelphia: Temple University Press.

Bookman, A., & Morgen, S. (1988). Rethinking women and politics. In A. Bookman & S. Morgen (Eds.), *Women and the politics of empowerment* (pp. 3-32). Philadelphia: Temple University Press.

Boston University. (1988). *Boston University's report on the Chelsea Public Schools*. Boston, MA: Author.

Castells, M. (1983). *The city and the grassroots*. Berkeley: University of California Press.

Crenson, M. (1983). *Neighborhood politics*. Cambridge, MA: Harvard University Press.

DiMaggio, P., & Mohr, J. (1985). Cultural capital, educational attainment, and marital selection. *American Journal of Sociology, 90*, 1231-1261.

Fischer, C. S. (1982). *To dwell among friends*. Chicago: University of Chicago Press.

Gottlieb, N. (1992). Empowerment, political analyses, and services for women. In Y. Hasenfeld (Ed.), *Human services as complex organizations* (pp. 301-319). Beverly Hills, CA: Sage.

Greer, S. (1962). *The emerging city: Myth and reality*. New York: Free Press.

Hardy-Fanta, C. (1993). *Latina politics, Latino politics: Gender, culture and political participation in Boston*. Philadelphia: Temple University Press.

Kennedy, M., & Stone, M. (1990). *The Hispanics of Chelsea: Who are they?* Boston: University of Massachusetts, Center for Community Planning.

Lawson, R., & Barton, S. E. (1990). Sex roles in social movements: A case study of the tenant movement in New York city. In G. West & R. L. Blumberg (Eds.), *Women and social protest* (pp. 41-56). New York: Oxford University Press.

Ley, D., & Mercer, J. (1980). Locational conflict and the politics of consumption. *Economic Geography, 56*, 89-109.

Lin, N., Simeone, R., Ensel, W. M., & Kuo, W. (1979). Social support, stressful life events, and illness: A model and an empirical test. *Journal of Health and Social Behavior, 20*, 108-119.

Lind, A. C. (1992). Power, gender, and development: Popular women's organizations and the politics of needs in Ecuador. In A. Escobar & S. Alvarez (Eds.), *The making of social movements in Latin America: Identity, strategy, and democracy* (pp. 134-149). Boulder, CO: Westview.

Logan, J., & Molotch, H. (1987). *Urban fortunes: The political economy of place.* Berkeley: University of California Press.

Luttrell, W. (1988). The Edison School struggle: The reshaping of working-class education and women's consciousness. In A. Bookman & S. Morgen (Eds.), *Women and the politics of empowerment* (pp. 136-158). Philadelphia: Temple University Press.

Mayfield, L. (1991). Early parenthood among low-income adolescent girls. In R. Staples (Ed.), *The Black family* (pp. 227-239). Belmont, CA: Wadsworth.

Massey D., & Denton, N. (1988). Suburbanization and segregation in U.S. metropolitan areas. *American Journal of Sociology, 94*, 592-626.

McCourt, K. (1977). *Working-class women and grass-roots politics.* Bloomington: Indiana University Press.

Morgen, S. (1988). "It's the whole power of the city against us": The development of political consciousness in a women's health care coalition. In A. Bookman & S. Morgen (Eds.), *Women and the politics of empowerment* (pp. 97-115). Philadelphia: Temple University Press.

Naples, N. (1991). Contradictions in the gender subtext of the war on poverty: The community work and resistance of women from low income communities. *Social Problems, 38*(3), 316-332.

Park, R., Burgess, E., & McKenzie, R. D. (Eds.). (1925). *The city.* Chicago: University of Chicago Press.

Pope, J. (1990). Women in the welfare rights struggle: The Brooklyn welfare action council. In G. West & R. L. Blumberg (Eds.), *Women and social protest* (pp. 57-74). New York: Oxford University Press.

Sacks, K. (1988). *Caring by the hour: Women, work and organizing at Duke Medical Center.* Urbana: University of Illinois Press.

Saunders, P. (1981). *Social theory and the urban question.* London: Hutchinson.

Stephen, L. (1992). Women in Mexico's popular movements: Survival strategies against ecological and economic impoverishment. *Latin American Perspectives, 19*(1), 73-96.

Susser, I. (1988). Working-class women, social protest, and changing ideologies. In A. Bookman & S. Morgen (Eds.), *Women and the politics of empowerment* (pp. 257-271). Philadelphia: Temple University Press.

Wellman, B. (1979). The community question: The intimate networks of East Yorkers. *American Journal of Sociology, 84*(5), 1201-1231.

West G., & Blumberg, R. L. (1990). Reconstructing social protest from a feminist perspective In G. West & R. L. Blumberg (Eds.), *Women and social protest* (pp. 3-35). New York: Oxford University Press.

Wilson, W. J. (1987). *The truly disadvantaged: The inner city, the underclass, and public policy.* Chicago: University of Chicago Press.

Wolfe, J., & Stracham, G. (1988). Practical idealism: Women in urban reform, Julia Drummond and the Montreal Park and Playground Association. In C. Andrew & B. Moore Milroy (Eds.), *Life spaces: Gender, household, employment* (pp. 65-80). Vancouver: University of British Columbia Press.

Zavella, R. (1987). *Women's work and Chicano families: Cannery workers of the Santa Clara Valley.* Ithaca, NY: Cornell University Press.

Part II

Politics

6

Getting Women's Issues on the Municipal Agenda: Violence Against Women

CAROLINE ANDREW

Municipal government has not been a feminist issue. This level of government has, to say the least, not been the principal target of the women's movement during the 1970s and 1980s, the second wave of modern feminism. National governments and, in the Canadian context, provincial governments were the principal foci for attention and the most active initiators of government policy. Women's issues gained a place on the federal and provincial policy agendas through a complex network of interactions between women's groups, government officials, public opinion, and elected politicians (Adamson, Briskin, & McPhail, 1988; Burt, Code, & Dorney, 1988). The local political agenda remained, up until the late 1980s, almost totally impervious to the social and political changes brought about by the women's movement.

This chapter evaluates recent Canadian municipal experience in the area of initiatives relating to women's issues, to determine the significance of these initiatives. This will be done, first, by examining a number of these initiatives and, second, through a case study of one such example, the creation of the Women's Action Centre Against Violence (Ottawa-Carleton, Ontario).[1]

AUTHOR'S NOTE: Most of my research on the Women's Action Centre Against Violence has been done with two colleagues from Carleton University: Fran Klodawsky (women's studies and geography) and Colleen Lundy (social work). I should like to acknowledge their influence on my thinking about the center and thank them for our intense discussions and for the personal and intellectual warmth and concern they both demonstrate in our attempts to understand the intricacy of the interrelationships between community and state. Thanks also to the wonderful waitresses at the Avenue Café, who cheerfully endure our endless meetings and equally endless requests for more coffee.

■ Framework for Analysis

Recent studies of municipal involvement in women's issues, largely from the United Kingdom and Canada, identify factors that appear to have played a significant role in efforts to increase the saliency of women's issues on the agenda of municipal government. One principal factor is the state of organization of the women's movement at the local level and its degree of mobilization. Municipal action has been the result of pressure from women's groups, and the precise form of this action has often been the direct result of the groups' political activity. Even those for whom municipal action is basically understandable in terms of the general pattern of gender relations in the community argue that the impact of this overall community structure is actualized through the political activity of the women's movement:

> Evidence suggests that the reason why only certain local authorities chose to set up women's initiatives arises from variations in local gender relations, which in turn are connected to variations in feminist political activity and ultimately in distinctive local policy outcomes. (Brownhill & Halford, 1990, p. 411)

Joyce Gelb's (1989) comparative analysis of feminism and political action in the United Kingdom, the United States, and Sweden insists on the link between the relative permeability of different levels of government and the structure of the women's movement in the different countries. She argues that the "decentralized structure of British feminism" has led to "a variety of multifaceted action groups" that have "turned to local council governments for funding, access and space" (p. 85). According to her analysis, the relatively closed national political system in Britain has offered fewer possibilities for the women's movement at this level; "political opportunity structures" (p. 221) have localized both feminist activity and agendas more than in the United States and Sweden.

Formal and Informal Politics

The form and composition of the local political structures also shape municipal initiatives. Whitzman (1992) argues that the terms of reference of the first task force by the regional government of Metropolitan

Toronto "were a result of lobbying by women's advocacy groups as well as the presence of 'femocrats' in local government" (p. 171). Wekerle's (1991) summary of the British experience emphasizes factors related to municipal structures and functioning, such as the political composition of councils and the number of female elected officials. Indeed, she argues that women's initiatives "have tended to be generated by municipal staff and politicians" (p. 2). Other factors that have been identified as influential in the British case were central government policies to promote economic and social regeneration in the inner cities and the sense of initiative of local planners (Trench & Jones, 1990; Trench, Oc, & Tiesdell, 1991).

However, rather than treating these two factors—feminist mobilizing and state structures—as two separate and perhaps even contradictory phenomena, Brownhill and Halford (1990) make an important methodological point in asking, "How useful is a formal/informal dichotomy?" for the understanding of women's involvement in local politics. Their answer is clear:

> Women in supposedly separate spheres face similar problems in their political organizing—for example, blockages to the achievement of effective results—but also . . . positive aspects of the process of women's organizing, and in particular the political consciousness and personal confidence that can flow from this, occur in both "spheres." Further, . . . the inter-connections which exist between political activity in the two spheres are so great and so complex that it is very difficult, other than at a most superficial level, to decisively allocate any form of political activity into one sphere or the other. This latter point is equally valid at both a practical level, where people are continually meeting, consulting or working jointly and where resources may change hands, as well as at a more conceptual level. (p. 398)

The authors are not arguing that there are not two levels of activity but, rather, that the links between local feminist organizing and local state initiatives as well as processes common to the two are more important than conceptual compartmentalization. Grasping the sense of the movement between grassroots activity and state activity or, as Brownhill and Halford (1990) state, being able "to challenge and transform such categories as the informal and the formal to tease out the political processes which underlie them" (p. 412) is significant to understanding, for example, the role of elected officials.

Political Discourse

The studies also identify the importance of the ideological framework or the "universe of political discourse" (Jenson, 1986). The ideas that are accepted at the municipal level justify action or inaction in different policy areas. The universe of political discourse at any one time defines the range of legitimate activities of municipal government and therefore influences not only the kinds of activity but also the categories of legitimate actors. Saggar (1991) has looked at this question in relation to the changing agenda of race relations issues in local government in Britain, where "differences and conflicts over race policy are often as much to do with rival interpretations of the underlying framework as they are to do with immediate policy options" (p. 101). For instance, the prevailing liberal policy framework "placed a heavy premium on assumptions of 'colour blindness,' " and, often, demands of groups were rejected "because of a broad predisposition on the part of local government members and officers to reject the legitimacy of such sectional interests" (p. 101). Conceptions about what are legitimate actions or actors and, moreover, what is the responsibility of local government are extremely important in guiding and shaping local policy.

One way of getting at this ideological dimension is to analyze the changing nature of the universe of political discourse and the emerging discourses and themes in local politics. This allows an examination of ideas, placed within a context of political struggle. De Leonardis (1992) talks about the new culture of rights and services, where social needs assume the status of rights and of active citizenship, and Fraser (1989) speaks of the "politics of need interpretation." For Fraser, "needs talk appears as a site of struggle where groups with unequal discursive (and nondiscursive) resources compete to establish as hegemonic their respective interpretations of legitimate social needs" (p. 166). These authors are clear that discourse is critical to changes in local politics. Fraser's "reason for focusing on discourses and interpretation is to bring into view the contextual and contested character of needs claims" (p. 163), and, indeed, she argues that issues become political only when they are "contested across a range of different discursive arenas and among a range of discourse publics" (p. 167).

The discourses about women's needs and their legitimacy at the municipal level explain how women's issues are, or are not, gaining a place on the political agenda. Fraser (1989) argues that "by speaking publicly the heretofore unspeakable, . . . feminist women have become 'women' in the sense of a discursively self-constituted political collectivity, albeit a very heterogeneous and fractured one" (p. 172). But she hastens to add that these oppositional discourses do not necessarily go unchallenged and one must examine in specific cases what is the result of contending discourses.

These studies suggest a methodological framework for examining whether the municipal government agenda has been influenced by women's issues. This methodology involves, first, looking at municipal structures, women's organizational mobilization and the dynamics of the complex relationships between them and, second, looking at local political discourses, as they touch on questions of the legitimacy of action and of municipal responsibility for women's issues. The framework elucidates the variety of recent initiatives in Canadian municipalities, and it provides a context for exploring in greater detail the example of Ottawa-Carleton.

■ Canadian Municipalities and the Women's Movement

Canadian municipalities have, for the most part, been closed to women's issues. Their principal activities have been related to the planning and provision of those major infrastructures necessary for the development of urban land (Higgins, 1986; Sancton, 1991; Tindal & Tindal, 1990). In the 19th century, these activities were accompanied by some municipal role in health and welfare services (the combination of these activities made municipalities significant political actors in the 19th century), but these responsibilities were given up to, taken up by, and/or lost to provincial governments during the first part of the 20th century (Taylor, 1984).

By the mid-20th century, Canadian municipalities had become relatively insignificant political actors, dominated by their provincial governments, which had centralized responsibility for most of the principal welfare state functions—education, health and social welfare. For the most part, municipalities tended to react to provincial decisions, rather than initiate their own activities.

In addition, Canadian municipal councils tend not to be organized along political party lines. The major exceptions to this have been Vancouver, Montreal, and Quebec City where there has been a tradition of civic parties, but even here few have evolved party systems with both majority and opposition parties. More typically, elected officials are chosen without party identification, and this has worked to fragment the potential for municipal leadership and direction (Higgins, 1986; Tindal & Tindal, 1990).

Recently, there are some indications that changes in activities and structures may be emerging that will modify these characteristics of Canadian municipalities. Municipal governments are beginning to act in social policy areas and are therefore broadening their base of activities from the very focused emphasis on infrastructures. At the same time, and interrelated to these changes in activity, there are changes in the social composition of municipal elected officials. Of particular significance here has been a recent increase in the number of women elected at the municipal level, particularly in the largest cities (Maillé, 1990).

In terms of the women's movement, the 1970s and 1980s saw an enormous development of service organizations, including transition houses for battered women, sexual assault support centers, and rape crisis centers. There has been much debate over the political significance of this evolution (Adamson et al., 1988; Lamoureux, 1990; Michaud, 1992), as to whether it represents or leads to co-optation by the state and therefore a weakening of the transformatory potential of the women's movement. Other authors (Masson & Tremblay, 1993; Ursel, 1990) argue that it is a beneficial move, showing the ability of the women's movement to effect real social change. Certainly, all agree that feminist services have greatly increased in number. Most of these organizations are local in orientation, and the greatly increased number of local frontline service organizations run by women for women creates increased possibilities for contacts with local governments. The needs of these organizations, for funding in addition to space, recognition, and services, can lead to their looking to municipal government.

Up until the present, the women's movement in Canada has not focused much attention or pressure on the municipal level of government. The principal demands of the women's movement were directed to the federal or to provincial governments, and maintenance of strong programs at the federal and provincial levels has been one of the ongoing demands of the women's movement. Decentralization has traditionally been seen as akin to privatization—as changes that were likely to put into question the gains made by the women's movement. This neglect and/or suspicion of local government is beginning to change, but it would certainly be false to say that strong ties exist between the organized women's movement and local government. That changes are occurring is linked both to the greatly increased number of locally oriented feminist services and also to the significance of the increasing number of women that have been elected to local councils and, indeed, of the increasing number of feminists among them.

Turning now to our second dimension, the analysis of the political discourses within Canadian local government in general, three strands can be seen as particularly pertinent to our analysis. The first is the discourse of limited local responsibility in the area of social policies and, particularly, redistributive policies. This is not a new discourse, but it has been refined and sharpened in the budget-conscious 1980s. It is in part an argument against taking on new areas of activity but justified with reference to the nature of the local tax base (inconsistent with redistributive policies, according to the argument) and to the particular form of the Canadian welfare state (defined in terms of clear provincial control over the areas of health and social services).

The second important theme in local political discourses is that of the closeness to the population. One of the favored themes of municipal actors is to argue that municipal government distinguishes itself from other levels of government by its very close links to the population. That this theme is, in fact, contradictory to the first does not prevent municipal politicians from using both of them, often at the same time. This theme is important in that it offers a space for the politics of needs interpretation, as Fraser (1989) suggests. If groups and individuals can define themselves as having certain needs and as being part of the local community, their claim on local government resources can be based on the argument that local government draws its legitimacy from its intimate links to the local population. This is therefore a discourse that can be used to make claims on local government.

The third element relates to the porousness of certain policy areas to a gendered analysis. There are a number of policy areas through which the introduction of women's issues at the municipal level might be done. The ones most often referred to in Canada have been planning (City of Montreal, 1989), housing (Mondor, 1990), equal pay (Findlay, 1993), and violence against women (City of Toronto, 1991; Dartmouth Task Force, 1992; Urban Safety, 1991; Women and Urban Safety Committee, 1991). At the present time, the issue of violence against women would appear to be the principal vehicle for making an argument in favor of municipal responsibility and municipal action.

The reasons for focusing on violence against women are numerous. It is an issue seen to be central by the women's movement throughout the 1980s and 1990s, an issue so crucial that action is needed by all levels of government and all sectors of society. As the Dartmouth Task Force (1992) stated,

> Because this issue was seen to impact greatly not only on the health of the individuals involved and their family members but on the integrity and functioning of the support systems in the larger community, there was overwhelming agreement that Dartmouth should take action. (p. 1)

Second, the issue of violence against women can be easily related to questions of urban safety or crime prevention, areas where the legitimacy of municipal government action has long been established. Indeed, both in the United Kingdom (Trench & Jones, 1990) and in Canada (Andrew, 1994), central governments have financially supported municipal activity in the area of crime prevention, arguing for both the importance of these activities for "economic enterprise and community life" (Trench & Jones, 1990, p. 1) and for the appropriateness of the municipal role. The link from general crime prevention activities to those relating specially to women has been made in a number of ways, by emphasizing the particular vulnerability of women (p. 2) and the extent to which research indicates that women curtail their activities "by fear of attack or harassment" (p. 3). There is therefore both a needs and a rights argument about the importance of looking at safety issues in terms of a gendered analysis and therefore of integrating violence against women into ongoing municipal activity on crime prevention. Both these arguments were expressed in the report of a British National Conference on Preventing Crime Against Women:

> As workers, carers, individual and community members, women's freedom to fully contribute to and share in social, economic and political life is restricted by crime against them. The direct and indirect costs of crime against women on health, social welfare and criminal justice agencies are considerable. (Holder & Stafford, 1991, p. 3)

The link to urban safety also suggests additional reasons why this issue has been picked up by municipal governments. Urban safety is compatible with, and indeed can be seen as a part of, the economic development strategies pursued by municipal governments. Acting on this issue does not detract from the "main business" of municipal governments, all the more because in most cases the amount of financial support requested is relatively modest. In addition, violence against women is an issue with a very great mobilizing potential. It attracts multiclass participation; it crosses age lines and ethnic and linguistic divisions as well as linking service providers, traditional women's organizations, and feminist groups. Indeed, an important political calculation, it is both a women's issue and one that crosses sex lines.

There are two other reasons, more particular to Canada, why the policy area of violence against women has been seen as the best avenue for the entry of women's issues into the sphere of municipal political activity. One is the impact of the Toronto experience, where initiatives relating to women have existed since 1973 (with the creation of a Status of Women Task Force) and where those initiatives have heavily concentrated on activities relating to violence against women. The best known examples in Toronto are the creation of an advocacy agency, the Metro Action Committee on Public Violence Against Women and Children (METRAC), stemming from a regional government task force in 1982, and of the Safe City Committee of the City of Toronto, created in 1988 (Wekerle, 1991; Whitzman, 1992). The Toronto example and the written material available from METRAC and the Safe City Committee have been extremely important in raising consciousness about the issues and in proposing models for action. Their influence can be seen in all of the reports done across Canada.

In addition, the 1989 massacre of 14 women engineering students in Montreal[2] greatly increased the sensitivity of all sectors of Canadian society to the question of violence against women. The evolution of the public discourse following the Montreal massacre reinforced the sense that this was an issue for community action and the view that this action must be broad in scope, encompassing both public and private violence.

■ Recent Canadian Municipal Initiatives

Both in terms of the interrelationships between grassroots organizing and state action and of municipal discourse, the traditional reality of Canadian local government has not been receptive to women's issues, but this traditional reality would appear to be in the process of change. These processes of change open up spaces of opportunity for new initiatives, actors, and discourses. Recent reports from Winnipeg, Manitoba, and Dartmouth, Nova Scotia, as well as Toronto's continuing activity, illustrate the politics of these new spaces. An analysis of these

reports permits a richer understanding of the political processes underlying the relationships between municipal governments and the women's
movement.

A number of dimensions are particularly clear. First of all, the role
of community groups and women's groups, both in doing the reports
and in influencing the kinds of actions being called for, is crucial.
Although the nature of the involvement of the women's movement
varied across the communities, reflecting the state of organization and
specialization of the women's movement and of community-based politics in general, in all cases this involvement was present. In turn, the
composition of the groups responsible for the reports related to the kind
of proposals made and to the exact nature of the balance between
community action and state action. In all reports, both state and community were present, but the balance between the two was different.
Toronto, the largest Canadian metropolitan center with a large, active,
and highly diverse women's movement, Winnipeg, a major regional
center with a tradition of community activism, and Dartmouth, a relatively small community in Eastern Canada situated next to Halifax, a
much larger center with an active, community-based women's movement, produced very different models of state-community relations.

In Toronto, the report, *A Safer City* (City of Toronto, 1991), was the
product of the Safe City Committee, composed of four members of city
council and of representatives from 14 community-based groups whose
mandate included the prevention of violence against women. The priority recommendation in Toronto's report was to increase community
resources to prevent violence against women through municipal financial support:

> Particular resource needs identified in the consultations included: devel-
> oping and translating culturally sensitive written materials, assisting in
> the formation and continuation of support groups for women surviving
> violence, and allowing community groups to develop and expand their
> counselling services. The exact criteria should be set in further consulta-
> tion with community-based groups and agencies. (p. 27)

The choice of a community-based model is very strong in the Toronto
report, and a certain distrust of state action is apparent in the underlying
principles of the report: "The City of Toronto should support non-
profit/community agencies working on the prevention of violence against
women, rather than developing parallel responses, or attempting to control

community initiatives" (p. 16). Municipal government is seen as a focal point for community action, but this action must remain firmly rooted in the community. The formal/informal dichotomy must disappear, and municipal resources must support and enhance the community network.

The Winnipeg study was conducted by a fact-finding group of 26 people, predominantly women, coming from the women's movement, community groups, and municipal services. The community input was strong but less specifically related to women's organizations working in the field of violence against women. The principle of community expertise was also recognized:

> The problem is everyone's, so too is the solution. These partnerships must include the commitment of all sectors of the community including individual citizens, City council, service providers, police, business and labour. It is an integrated approach which will require patience and creativity and one which must be guided by the voices and experiences of women. It is an approach based in community empowerment with the ultimate intent of creating a safer city for all. (Urban Safety, 1991, p. vii)

The Dartmouth Task Force was composed of 15 people, two thirds of them women. The members represented "the residential, business, and professional community, including legal, municipal, health, education, community services, minority and linguistic sectors. . . . An effort was made to assemble a collection of individuals already involved in the community and, if possible, with representative knowledge about violence against women" (Dartmouth Task Force, 1992, p. 52). The main recommendation was for the establishment of a committee with strong community representation but also including representatives of the three levels of government to "act as a focal point and springboard for community action" (p. 5). The Dartmouth report recommended a balance of state and community action, but the balance was more toward state action than was that of Winnipeg or Toronto.

The form and strength of the women's movement therefore influenced the models of action that were recommended. Where the local women's movement was strong, community solutions were preferred. This impact of the nature of the local women's movement, however, must be seen in interaction with the nature and structures of the municipal state. Dartmouth has been very active in the Healthy Communities initiatives [3] and for this reason was committed to municipal action and, in addition, very interested in horizontal coordination across policy sectors. Toronto had already created a specialized structure within the municipal bureaucracy, and the report was concerned to enhance this structure as the appropriate body to relate to the community. Specifically, the report recommended adding a staff person "to assist in developing communities' responses to increasing violence in Toronto and to provide support to existing community-based groups and agencies" (City of Toronto, 1991, p. 24). Winnipeg's municipal council and bureaucracy were less involved in their report, and this relates to the lesser degree of municipal government initiative recommended in the report. The balance of state and community action represents a choice, and, obviously, this choice is influenced by the capacity and will of community groups and by the capacity and will of the municipal structures.

The arguments for municipal action made in these reports have certain common themes. The basic thrust is double: There is an urgent need to act, and in a dramatic fashion, but at the same time this action follows the path of traditional municipal activity and is therefore legitimate. The sense of urgency is expressed differently in the reports. Winnipeg makes explicit reference to the victims of the Montreal massacre and continues: "We can no longer sit safe and secure in the knowledge that violence is something that happens to others by others. Violence is here. In Winnipeg" (Urban Safety, 1991, p. vi). In Dartmouth, the urgency related to the impact of violence on women ("Since November of 1988, 13 women have been murdered in Nova Scotia" [Dartmouth Task Force, 1992, p. 1]) but also to the intergovernmental context ("An immediate action plan is imperative if this City is to reap the benefits of the current economic and national focus on violence against women" [p. 3]). Toronto's report makes a triple argument for rapid action: maintaining Toronto's "groundbreaking role" (City of Toronto, 1991, p. 11), the financial limitations of community groups, and "exploring—and beginning to stem the root causes of violence against women" (p. 17).

Along with the urgency, the reports also indicate that important areas of traditional municipal responsibility are clearly linked to issues of violence against women. The capacity of the police to intervene effectively in cases of violence against women; the role of municipal planning to better integrate safety concerns into the physical design of cities, parks, and recreation planning that gives greater priority to women's safety; financial support to community groups; offering a vehicle for local coordination; and networking of groups involved in actions against violence against women—all of these touch on fully legitimate areas, or types, of municipal activity.

In addition, the reports all emphasize the importance of looking at the variety of women and women's groups. Violence against women should not suggest a homogeneous group; the heterogeneity of women is a crucial principle of action. For this reason, Toronto's report recommends "a priority given to services for immigrant and visible minority women" (City of Toronto, 1991, p. 26). This concern is shared across all the reports—adolescent women have particular needs, as do Native women, poor women, women with disabilities, lesbian women, elderly women. This recognition of diversity has important implications for government action. It reinforces the importance of community input and support for community activity because it is the total network of community groups that reflect the diversity of women. As the Winnipeg report argues, "Within Winnipeg and across the country, the voices and experiences of women must continue to guide the signs of change" (Urban Safety, 1991, p. 21).

These reports therefore make a particular kind of argument in calling for municipal activity. They argue that municipal government has a responsibility to act in this area because of its role as spokesperson and focus for community concerns. However, municipalities must not act in isolation but, rather, as part of a community network. This is necessary to ensure that women's issues and, indeed, a feminist perspective are integrally connected to municipal policy making. The best way of doing this is by ensuring community involvement in the formulation of municipal policy.

The areas for action relate to traditional areas of municipal responsibility but extend these activities in the direction of a more established presence in the social policy field. At the same time, these reports underline the links between physical and social planning and therefore demonstrate the social policy implications of physical planning and development. The intervention proposed is related to traditionally legitimate areas of activity for municipal governments—new actions are proposed, and proposed urgently, but argued in terms of activity understood to be within municipal jurisdiction and competence.

■ Women's Action Centre Against Violence— A Case Study

To understand the influence of structures—both that of the women's movement and of municipal government—and of discourse in local initiatives to put women's issues on the municipal agenda, let us look at the creation of the Women's Action Centre Against Violence (Ottawa-Carleton). Very briefly, the creation of the center represents a 2-year organizing process on the part of a broad community-based network of women's groups, service agencies, and individuals. The group met first in May 1990, and in April 1992 the Regional Municipality of Ottawa-Carleton agreed to initial funding. The center is currently operating, with program activities in the areas of safety audits, workshops for area planners, compiling a handbook on available educational resources, and doing public education.

The focus here is on the decision related to financial support, because it was this decision of the Regional Municipality that permitted the creation of the center and that represents an acceptance of a local government responsibility in the area of the prevention of violence against women. Looking first at the relationship between community organizing and state initiative, the creation of the Women's Action Centre illustrates admirably the point made by Brownhill and Halford (1990), that the dichotomizing of formal and informal sectors should be avoided. The crucial elements in setting up the center were a grassroots community-based network and an elected representative who initiated and directed the network. To see her role as being only part of the governmental structures is as erroneous as to see it as an integral part of the community network. It is both and perhaps something else again.

The initiative to do something in Ottawa started in early 1990 when Councilor Diane Holmes went to an international conference on crime prevention, in Montreal (Andrew, Klodawsky, & Lundy, in press). The conference was important for her in two ways: She learned more about the activities of Toronto's METRAC, and she listened to presentations that argued that the issue of women's safety is not only an issue of the fear of violence but even more an issue of the prevalence of violence. The combination of these factors convinced her that something should be done but also, and more important politically, that something could be done. She returned to Ottawa hoping to find an existing women's group that would take up the challenge of lobbying to create an organization in Ottawa similar to METRAC. After contacting a number of groups who were busy with other priorities, she realized that she would have to create an organization to do the lobbying.

In May 1990, she called together a small group of women, and from this point on the network grew. Two community forums were organized, to raise consciousness and test community support for the general direction in which the network was developing. During this period, the group met at least monthly, with subcommittees meeting more frequently to complete specific tasks. As well as organizing the forums, the group prepared a brief for the regional council outlining the need for a specialized agency and supporting their position with data demonstrating the extent of violence against women in Ottawa-Carleton. These data were compiled in part from police statistics, frontline service agencies, and a newspaper survey by the major Ottawa newspaper, *The Citizen.* In addition, the group organized support letters from a wide variety of community groups.

The groups, organizations, and individuals that did this work represented a broadly based community response to the issue of violence against women. These were representatives from traditional women's groups (for instance, the Business and Professional Women), frontline service organizations—both services specifically for women (Sexual Assault Support Centre, Rape Crisis Centre, transition houses) and more general community services (community health centers). There were representatives from the area universities, the school boards, the public transit commission, and the police. People were there from men's groups, senior citizen's groups, and local political activists. Some parts of the community were not represented well—the French-speaking community, the immigrant and visible minority community, and women with disabilities were the most serious omissions.[4]

There were some concerns expressed by frontline service organizations that support for the Women's Action Centre might mean less funding for frontline services (and more of these concerns have emerged since the beginning of the operations of the center), but for the most part, groups argued that increasing local government responsibility in this area could only increase public awareness of the seriousness of the issue and of the importance of public action.

The network was held together in part because of the focus on concrete activity and a very clear goal—getting regional government to fund the center. This direction was very strongly imprinted on the activities of the group by its chairperson, Councilor Diane Holmes. This brings us back to the interrelation of community mobilization and state action, because her actions in keeping the group focused on its task related in part to her role as an elected official (keeping the group aware of political realities) and in part to community mobilizing (continual pressure to keep the group as broadly based as possible). She was seen both as part of the municipal government and as part of the community effort to influence the government.

The group's efforts to mobilize community support can be seen as a delicate, and uneasy, balance between two strategies of change: "disengagement (the desire to create alternative structures and ideologies based on a critique of the system and a standpoint outside of it) and mainstreaming (desire to reach out to the majority of the population with practical feminist solutions to particular issues)" (Adamson et al., 1988, p. 23). These were not so much explicit choices but, rather, a series of decisions that aimed sometimes for positive goals (autonomy, reaching the population) and sometimes for avoiding the perceived dangers of marginalization (the negative side of disengagement) or of co-optation (the negative side of mainstreaming) (Andrew et al., in press).

The outcome was also influenced by the structures and composition of the regional government. The group's initial request was sent to the Planning Department for a staff report, and the existence of sympathetic staff meant that the report provided the kind of information that was needed to lead to positive support. The initial recommendation attached to the report by the regional executive committee was to argue that this was not an area of responsibility (but that it would give the funding), but this initial recommendation was replaced by one indicating that it was an area of responsibility for regional government (and that the funding would be given). The factors that appear to have been the most influential were the publicity from the newspaper survey (as well as a good deal of general newspaper coverage of the issue) and the broad base of the network. Regional politicians seemed to feel that "their" community had spoken clearly and that this could not be ignored. If they were not necessarily convinced that violence against women was a regional government issue, they were convinced that the particular request should not be ignored. By combining media attention and letters of support from community groups, the organizing effort was able to convince the regional politicians that a lack of support would be clearly visible to their electorate.

It remains to examine the discourses of the Women's Action Centre and to see how these elucidate the process of introducing women's issues at the municipal level. Two central themes can be identified, and both should be understood in terms of the broad and varied base of the network. Both represent themes that brought together the various parts of the network and also established the terms of their interaction with the municipal level of government.

The two principal discourses were that only a feminist analysis of violence against women could produce effective policy directions and, second, that municipal governments have responsibilities in areas that are significant for the prevention of violence against women. The first theme centered around the conviction that only by understanding violence against women as being rooted in the unequal power relations of society could effective policies be implemented. This theme was extremely important to the cohesion of the network, by focusing on what brought the participants together and on what united them in opposition to those social forces that do not recognize the policy pertinence of feminism. Moreover, this theme introduced a way of going beyond the division between "public" and "private" violence, therefore allowing equal participation by the frontline service organizations (whose work mostly related to private violence) along with the original participants in the network, whose background tended to be more in the area of public violence.

Putting forward the pertinence of a feminist analysis of violence as the central focus to the network also allowed for the legitimate participation of a wide variety of groups and individuals. The police and the public transit commission could be present, without difficulty, because their representatives could agree to the usefulness of this kind of analysis. Men could participate, and membership could be both as individuals or as representatives of groups or organizations.

At the same time, this theme in the group's discourse related to the strategy of disengagement by arguing that the community network's knowledge and expertise in this area was superior to that of regional government. The community should be empowered to act—with regional government resources.

The second major theme in the discourse is more allied to a mainstreaming strategy, in arguing that municipal governments have important responsibilities in areas relating to violence against women. This theme also arises from a compromise or a synthesis of positions coming from different parts of the network. For some members of the network, the absolute priority is the issue of violence against women, and the level of action—local, national, international—is largely irrelevant. For others, the primary interest is women's overall presence in municipal politics, and these people are less concerned about the particular avenue by which this objective could be realized. But both these points of view, and the groups and individuals holding them, could agree to put forward the argument that municipal government has important responsibilities in the area of violence against women.

This element in the overall discourse of the Women's Action Centre can be examined in terms of its practical consequences for the group, in keeping it focused on the specific tasks of building an effective lobby to influence regional council. At the same time, it also articulated a position that could be shared by those in local government. As was seen earlier in the more general discussion of recent Canadian initiatives, it is important to tie new demands into traditional areas of local government activity.

The theme of municipal responsibility more directly engages municipal government. It implies a criticism of the existing situation, but it expresses this in terms of an important potential for municipal activity and municipal responsibility. It plays back to the question of municipal activity in areas such as crime prevention and urban safety and argues that this activity should be developed, within a gendered understanding. It is on this point that the two themes come together—a feminist understanding and a municipal responsibility.

■ Concluding Remarks

The focus of this chapter is an analysis of the permeability of municipal government to women's issues. This has been done through an examination of recent Canadian municipal initiatives with particular attention to one example, the Women's Action Centre Against Violence (Ottawa-Carleton). Some success has been achieved in getting the local level of government, traditionally very closed to women's issues, to recognize some responsibility for the reduction of violence against women. The text has been centered on trying to understand how these successes have been achieved, and the framework used has looked at both the interrelation between feminist organizing and government action and the production of discourse. At the same time, one must not overemphasize the degree of success. At the most, these conclusions should take up the traditional analogy of the glass that, depending on one's perspective, can be seen to be half-full or half-empty. One can focus on the success of certain examples, or one can focus on the overall reality of municipal government, still eminently capable of ignoring gender. What status should be given to the successes? Are they minor exceptions or indications of an emerging future?

NOTES

1. I should state that my research on the center has been from the position of a participant-observer. I was part of the organizing group from late 1990 and am currently a member of the board of the center.

2. On December 6, 1989, 14 women engineering students were killed by an armed assassin who entered the engineering school, l'École polytechnique, and shot down women, declaring his hatred of feminists.

3. The Healthy Communities Project was the Canadian reaction to the ideas of the World Health Organization's "Health for all by the year 2000." It involved networking among municipally based initiatives that took very broad social-development approaches to health and that were concerned with interdepartmental and intersectoral strategies at the municipal level (see Andrew, 1994; Mathur, 1991).

4. It is extremely important to understand the processes by which these exclusions originated and persisted (see Lundy, Andrew, & Klodawsky, 1994).

REFERENCES

Adamson, N., Briskin, L., & McPhail, M. (1988). *Feminist organizing for change.* Toronto: Oxford University Press.

Andrew, C. (1994). Federal urban activity: Intergovernmental relations in an age of restraint. In F. Frisken (Ed.), *The evolving Canadian metropolis: A public policy perspective* (pp. 427-457). Berkeley and Toronto: IGS Press and Canadian Urban Institute.

Andrew, C., Klodawsky, F., & Lundy, C. (in press). Women's safety and the politics of transformation. *Women and Environments.*

Brownhill, S., & Halford, S. (1990). Understanding women's involvement in local politics: How useful is a formal/informal dichotomy? *Political Geography Quarterly, 9,* 396-414.

Burt, S., Code, L., & Dorney, L. (1988). *Changing patterns: Women in Canada.* Toronto: McClelland & Stewart.

City of Montreal. (1989). *Women and the city: Report of the Committee on the Problems of Women in an Urban Environment.* Montreal: Author.

City of Toronto. (1991). *A safer city: The second stage report of the Safe City Committee.* Toronto: Author.

Dartmouth Task Force. (1992). *Violence against women.* Dartmouth, Nova Scotia: Author.

De Leonardis, O. (1992, August). *New patterns of collective action in a post-welfare society: The Italian case.* Paper presented at a workshop on social policy, Aylmer, Quebec.

Findlay, S. (1993). Democratizing the local state. In G. Albo, D. Langille, & L. Panitch (Eds.), *A different kind of state?* (pp. 155-164). Toronto: Oxford University Press.

Fraser, N. (1989). *Unruly practices.* Minneapolis: University of Minnesota Press.

Gelb, J. (1989). *Feminism and politics: A comparative perspective.* Berkeley: University of California Press.

Higgins, D. (1986). *Local and urban politics in Canada.* Toronto: Gage.

Holder, R., & Stafford, J. (1991). *Crime against women.* London: London Borough of Hammersmith and Fulham and Crime Concern.

Jenson, J. (1986). Babies and the states. *Studies in Political Economy, 20,* 9-46.

Lamoureux, D. (1990). Les services féministes: de l'autonomie à l'extension de l'État-providence. *Nouvelles pratiques sociales, 3,* 33-44.

Lundy, C., Andrew, C., & Klodawsky, F. (1994). *From Coalition to Alliance: Organizing for social change in Ottawa-Carleton.* Unpublished manuscript.

Maillé, C. (1990). *Vers un nouveau pouvoir*. Ottawa, Ontario: Conseil consultatif canadien sur la situation de la femme.

Masson, D., & Tremblay, P. A. (1993). Mouvement des femmes et développement local. *Canadian Journal of Regional Science, 16*.

Mathur, B. (1991). *Perspectives on urban health*. Winnipeg, Manitoba: Institute for Urban Studies.

Michaud, J. (1992). The welfare state and the problem of counter-hegemonic responses within the women's movement. In William K. Carroll (Ed.), *Organizing dissent* (pp. 200-214). Toronto: Garamond.

Mondor, F. (1990). Le logement: Point d'ancrage pour un nouveau départ. *Canadian Women Studies, 11*, 46-47.

Saggar, S. (1991). The changing agenda of race issues in local government: The case of a London borough. *Political Studies, 39*, 100-121.

Sancton, A. (1991). The municipal role in the governance of Canadian cities. In T. Bunting & P. Filion (Eds.), *Canadian cities in transition* (pp. 462-486). Toronto: Oxford University Press.

Taylor, J. H. (1984). Urban autonomy in Canada: Its evolution and decline. In G. A. Stelter & A.F.J. Artibise (Eds.), *The Canadian city* (pp. 478-500). Ottawa, Ontario: Carleton University Press.

Tindal, C. R. & Tindal, S. N. (1990). *Local government in Canada* (3rd ed.). Toronto: McGraw-Hill Ryerson.

Trench, S., & Jones, S. (1990). *Planning for the safety of women in cities* (Nottingham Planning Working Papers, No. 4). Nottingham, UK: University of Nottingham.

Trench, S., Oc, T., & Tiesdell, S. A. (1991). *Safer cities for women* (Nottingham Planning Working Papers, No. 12). Nottingham, UK: University of Nottingham.

Urban Safety for Women and Children Project. (1991). *A safer Winnipeg for women and children*. Winnipeg, Manitoba: Author.

Ursel, J. (1990, November). *Considering the impact of the battered women's movement on the state: The example of Manitoba*. Paper presented at Women and the Canadian State, Ottawa, Ontario.

Wekerle, G. R. (1991, July). *Gender politics in local politics*. Paper presented at the annual meeting of the Association of Collegiate Schools of Planning, Oxford, UK.

Whitzman, C. (1992). Taking back planning: Promoting women's safety in public places—the Toronto experience. *Journal of Architecture and Planning Research, 9*, 169-179.

Women and Urban Safety Committee of Ottawa-Carleton. (1991). *Women and urban safety: A recommendation for action*. Ottawa, Ontario: Author.

7

Gender and the Politics of Affordable Housing

SUSAN ABRAMS BECK

Efforts to make women full participants in politics, not just as the governed but as the governors, have been based on the assumption that as more and more women assume office, important changes will be forthcoming. Indeed, as the number of women in office has grown, fundamental transformations have been realized.

Change occurs as new groups find their way into the decision-making bodies of government, either through the pressure process or by taking on a representative function. However, both of these routes reflect the idea that as women enter the political arena, they will do so on the terms already established, that is, on the terms that have been established by men. Still, there is considerable evidence that change will be real and far-reaching. New issues will be raised, new priorities will be established, and the political agenda will undergo a significant shift in the kinds of matters it considers (Dodson, 1991; Dodson & Carroll, 1991). This change is far more evident in national, state, and city government than in smaller communities. In the upper reaches of government, there is ample evidence that women have raised and persisted in addressing issues such as abortion, child care, parental leave, pregnancy leave, pay equity, educational equity, and employment opportunities. These cannot be brushed aside, for they are part of a general revolution in society's commitment to its citizens. However, on the local level, the introduction of women into the political arena leaves a more mixed result because there is less opportunity to raise these burning issues of the women's agenda. The matters that come before municipal governments rarely involve questions of pay equity or reproductive rights, although, clearly, there are some exceptions. Nevertheless, as the number of women on the local level has grown, there has been evidence of what Janet

119

Flammang (1984) has referred to as the "quiet revolution," (p. 10) a slow, but fundamental change in the substance and style of governance.

Research on women in local government has revealed that the distinctions between men and women are seen in the roles they perceive and act on, although substantively little has changed (see, e.g., Flammang, 1985; Johnson & Carroll, 1978; Lee, 1976; Merritt, 1980; Mezey, 1978a; Stanwick & Kleeman, 1983). When research concerns the constituency that women councilmembers should serve, it has focused on the women and men who have a certain agenda in mind, including such issues as the Equal Rights Amendment, pro-choice, and comparable worth.

Thus change requires women representatives to take action on behalf of these preferences, "acting for" women in a substantive way, not just fulfilling the descriptive representation by "standing for" them (Pitkin, 1967). In her examination of politicians in Hawaii and Connecticut, Mezey (1978b, 1978c) found that women's attitudes toward feminist policies were not more favorable than men's. Moreover, she found the issue priorities of men and women to be almost perfectly correlated, as did Merritt (1980). However, these studies focus on attitudes rather than actions and concentrate on women's policies as issues separate from the more traditional concerns of local governments.

The local community is the place where most citizens come up against what government is all about—garbage collection, streets plowed after a snowstorm, safety—and it is where most citizens have an opportunity to get to know their public officials. Generally, the local arena is not perceived as the locus for discussing policy issues of specific interest to women. Smaller towns usually do not deal directly with issues such as abortion. But it is true that the lives of women are deeply affected by the decisions made by local governors: decisions that touch where women live, the availability of child care, and how they will deal with difficulties in their senior years. Moreover, there are the everyday concerns of localities that might be treated differently by women than by men.

Analysis of changes at the local level is usually informed by an understanding of the liberal tradition in the practice of American politics. For the analyst, politics is conceptualized as interests straining against one another and government brokering the demands of one group against another. Officials operate according to the ideas inherent in liberal democratic thought, confining their understanding of representation within the paradigm of balancing conflicting interests. Thus change can be discussed only in terms of groups and interests. The liberal framework of government at this level inhibits discussion of issues that cannot be conceptualized as interests and treated by balancing the demands of one group against another.

Feminist theorists have argued that there is a relational way of understanding politics, one that is not constrained by the limits of liberal thought, and more capable of ensuring social justice (Disch, 1991). Such a model must take into account the differences between people who are shaped by the social contexts in which they live (e.g., Diamond & Hartsock, 1981; Hartsock, 1985; Minow, 1990; Young, 1990). Women, as Martha Ackelsberg (1984, pp. 249-250; 1988) writes, have always experienced a blurred line between the public and private worlds, negotiating with landlords and public agencies, school boards, and transportation bureaus, even as they have led supposedly private (domestic) lives. Conceptualizing these arenas as separate, either spatially or functionally, is an artifact that serves to isolate women, keeping their concerns off the agenda and discounting their perceptions.

A new form of politics recognizes the connections between the public and domestic life and seeks to validate the introduction of women's experiences into the public sphere. The analysis of the practice of politics on the local level, the most intimate level of politics, offers the possibility of seeing how women officeholders might be the agents of such change. Their life experiences may translate into approaching governmental decisions differently.

However, even on the local level this will not occur as long as women continue to operate by the rules and ideas that are already deeply established. For change to occur, the principles of individual rights, rationality, and universality will have to operate along with principles that emphasize experiences of individuals who are connected to each other. Thus the ability of women to make a difference depends not only on their ability to seize their "fair share" but to rethink how politics, particularly government, operates. It is not an argument to replace one with the other but to recognize the restrictions imposed by operating only within the given framework.

This article investigates the liberal and relational models of governance through a case study of affordable housing in five suburban communities. The municipalities examined are located in New Jersey. In 1983, the New Jersey Supreme Court handed down the *Mt. Laurel II* decision,[1] which held that every municipality in the state has an obligation to provide, through its land use regulation, a realistic opportunity for the construction of its fair share of low- and moderate-income housing. This decision declared exclusionary zoning to be unconstitutional and thus challenged the liberal framework on which local land use laws had traditionally been based. According to the decision, zoning ordinances have to include affirmative ways to attract affordable housing, such as mandatory set-asides for builders putting up market rate units and/or providing builders with bonuses or even zoning to permit mobile homes.

Following the *Mt. Laurel II* decision, there was an enormous hue and cry from communities throughout the state, particularly suburban localities, which argued that the Supreme Court had violated the principle of home rule and had unjustifiably intervened in the affairs of municipalities. In response to the uproar, the New Jersey legislature passed the Fair Housing Act of 1985, which established the Council on Affordable Housing (COAH), a state administrative panel, to calculate the obligation of each town and oversee the development of affordable housing plans.[2]

COAH's implementation of the Fair Housing Act softened the impact of *Mt. Laurel* on municipalities, reducing the number of affordable units that localities were assigned and shifting the goal from making housing more affordable to focusing on existing substandard housing. COAH's policies were, and continue to be, extremely controversial.

Thus the requirement for a fair share of affordable housing placed a policy problem on the agenda of all localities, providing a common ground on which to compare councilwomen's and councilmen's perceptions and decisions and the extent to which the different experiences of women have translated into "relational" politics. The following study, based on a series of interviews in five New Jersey towns,[3] indicates that women do practice relational politics on some affordable housing issues that deal directly with residents within their communities. However, women and men coalesce in the unified front they present to any external threats, and in these contexts their competitiveness and protectiveness are consistent with the liberal tradition.

■ Liberalism in Localities

In the liberal state, rational individuals pursue their own preferences and wants, and the state protects individuals by providing an orderly society in which each person is allowed autonomy and thus liberty.

There is a theoretical equality among all individuals, and the state's obligation is to maintain neutrality in relation to its citizens (Young, 1990, pp. 76-78). Rules are based on rationality, not feelings, because emotion leads to arbitrariness and, therefore, a lack of fairness.

As women—one group among many—enter the public arena that is distinguished from the private arena, they will take their place alongside others, learning to articulate their demands, press their interests, and gain positions of power when and where they can. They seek their share of power.

Local government that operates under such a system competes in the municipal marketplace with other cities and towns. Those who like the decisions of one town better than another are free to "buy" the product of that town by moving to it (Logan & Molotch, 1987). Towns that do the best job of providing services to their residents will be in highest demand, and consequently their product will cost more. Thus housing costs rise, and the value of living in one town surpasses that of living in another.

In an effort to gain a prominent place in the competitive market, ruling bodies in such communities must foster an attractive product. In the case of housing, and in particular revolutionizing zoning laws to accommodate low- and moderate-income housing, certain unwritten rules and values enhance a community's competitiveness. They involve (a) maintaining a low tax rate while keeping property values high, (b) protecting the sanctity of single-family homes, (c) minimizing development except where "clean" ratables (low-rise office buildings in bucolic settings) can be attracted, (d) working to establish or maintain an attractive "downtown," and (e) protecting the ideal of home rule at all costs. The unspoken subtext that runs through all these values is to keep inner-city poor minorities at a distance.

In the interviews, these themes were repeated over and over again by women and men on local councils, who cite taxes and development as the most important issue in the community: "If you let things run down, your values go down, your properties go down, your educational system goes down. You have to keep on top of it. I think that's what being a legislator in town is all about." Both Democrats and Republicans criticize the opposing party for being receptive to proposals for a hotel and for high-rise offices. Concern over the development and growth that did occur in one town "turned everything topsy-turvy. New people have come in from New York City and [a small city nearby] with a lot of money, and those people want to remain isolated." Every person interviewed mentioned the importance and difficulty of maintaining a low tax rate. And in every town, the development of the downtown was clearly in the front of their minds, as they worried about the "commercial areas that are under great stress, both because of financial pressures and because of the competing malls. . . . This ripples throughout the whole town: jobs, appearances, and ultimately property values."

Indeed, in one municipality, there was a change in the form of the government that resulted from the anti-high-rise sentiment. In response to a project that was only a few stories, "People went crazy" because they want "to escape density . . . [and] to see lots of lawn." It is, according to a councilman, a total devotion to the "sanctity of the single-family home" coupled with "a very strong antidevelopment sentiment, maybe to the extreme. . . . There's this perception that condominiums are undermining the rural character of our community."

Control over ecodevelopment issues in suburbia is anchored in the concept of home rule. Members of local councils—women and men—are extremely protective of this tradition and resent state interference in their affairs. For example, they expressed their anger over the 1.5% cap that the state had imposed on budget increases, the state's mandatory labor guidelines, and their deep displeasure over the state's *Mt. Laurel* demands.

Because women and men agree about the basic rules of governing, it is not surprising that they come to similar conclusions over implementing affordable housing. At the time of these interviews, the period of explicit resistance to *Mt. Laurel* had passed.[4] Even though all five communities had built some affordable housing, both councilwomen and councilmen expressed reservations about the decision and its implementation. At the extreme, it was called "socialism." They argued that "home rule had been thrown out" and that "you can't mix socialism and home rule. If there's a great need for affordable housing, the state should have provided money for it. . . . We rezoned, but only under protest." "What the state mandates, the state should pay [for]," commented one councilwoman. Less vehement, but still upset, another argued that "there is supposed to be local control, but there really isn't. It is an example of the state interfering with local prerogatives."

Probably the most common way that women and men members of councils have come to deal with *Mt. Laurel* is to reluctantly accept it but to think of it strictly in terms of the local population or the population in communities adjacent to them and like them. This justification permits them to support *Mt. Laurel* in principle but concomitantly reject the prospect of inner-city non-Whites moving to the suburbs. It is a powerful argument, because it permits them to recast *Mt. Laurel* in a way that makes it perfectly consistent with the already accepted suburban values, preventing outsiders from infringing on their communities' quality of life.

This argument was articulated by virtually all those interviewed, some more vigorously than others, but all had a similar view. The rationale includes a perception that their town is diverse and that therefore, to meet the needs of their own population, they must provide affordable housing. "You have people in town," states a councilman, "who fit that description of low income. You have people who are renting. You have DPW [Department of Public Works] workers. You have people who went to [our] schools and didn't go to college and have blue-collar jobs. They fit the low-income housing." Both women and men cite new immigrant groups that had settled in fairly large numbers within their borders, the mix of blue- and white-collar workers, the single-parent families, and the numbers of young people just starting out. But as the preceding statement implies, while emphasizing their own diversity, there is concurrently a rejection of "others," that is, those who represent the inner-city populations. One councilman argued, "We have enough people who need it in our region. People don't understand this. They think of *Mt. Laurel* as bringing in slums, but we need it for people who live right here." Another commented, "I think we ought to take care of our own. . . . I'm not looking to bring in low-income people from other areas and overwhelm [our town]."

Clearly, there is an understanding of the power of racial politics in this kind of thinking. One councilman admitted that the town's residents did not "feel much obligation to the cities," and to avoid creating "two New Jerseys, one suburban and one urban, we have to do something now." But he did not think "mandating affordable housing will work. . . . The political decision is to reserve it for your own community."[5] Fear of the cities was expressed by a councilwoman who admitted that she is uncomfortable with the idea of *Mt. Laurel* housing if it results in "mixing" the inner-city population with those in her town. Her co-councilwoman made a similar argument, agreeing that she would "support *Mt. Laurel* for those employed in the town," but she felt no obligation to residents of other communities, even if another town offered employment opportunities for the people in *her* community. "After all," she noted, "We wouldn't get any of the advantages of having a 'ratable' in our town, and we'd get all the aggravation: pollution, traffic, etc." A councilwoman in another community rationalized that "the low-income people, or the Black people, never wanted to come to the White school." She argued that housing a poor person "in the middle of a wonderful area" would make that person feel "stigmatized," out of place in their "rags." This rationale, which argues that connection is spurned by both suburban and city dweller alike, brings the liberal argument full circle.

Councilmembers also find other excuses to avoid full acceptance of *Mt. Laurel*. Supportive in principle, they then claim that they do not have any room; they have already built some, enough, or even too many units; their town already offers housing across a wide spectrum of income levels; they do not want to look like a neighboring, less attractive town; they cannot support the strain that such housing will put on their infrastructure, such as the need for more schools, increased traffic, or an increased need for social services. Indeed, every town examined in the entire sample had examples to support one or more of those arguments. These justifications permit officeholders to avoid facing the political consequences of the specter of integration. As would be expected, in the suburb that included a significant minority population, councilmembers faced a political landscape where race relations within the community were fragile. But even there, councilmembers did not articulate a sense of responsibility to residents in urban centers and remained focused on maintaining stability within their town.

Thus the overwhelming sense of their attitudes toward *Mt. Laurel* is that, although they support the idea of affordable housing, women and men on local councils are willing to address this need primarily for their own, or for those just like them, but they want as little as possible to do with a fair share for the urban, minority poor. There are some who sense they might like it to be otherwise—men and women—but they refuse to translate that attitude into any kind of action.

This attitude flies in the face of the spirit of the *Mt. Laurel* decision, as it was handed down by the New Jersey Supreme Court, but it is perfectly consistent with the Fair Housing Act, which reflected the suburban perspective. Affordable housing has been built in the suburbs for the suburbs, and that has done little to bridge the gap between city and suburb. Liberal governance by both women and men on local councils has maintained the status quo, thus severely constraining *Mt. Laurel*'s impact.

The question remains as to what extent councilwomen, who internalize liberal ecodevelopment norms, can change what local politics is all about. Do women officeholders engage in interest politics, perhaps adding a new set of issues to the local agenda, or, as many scholars have argued, do they conceptualize politics differently, relationally?

■ Relational Politics

A relational approach starts from the premise that knowledge derives from human experience, but such experience is socially constructed and is not some uniform, abstract humanness. In understanding political life through an individual's role as part of a community rather than as abstracted from that community, understandings are themselves human. One person cannot know—understand—the world in exactly the same way as another. There is no monolithic perspective from which all individuals see their connections to others. Neutrality is a pretense, because any given "version of reality" (Minow, 1990, p. 389) reflects a certain perspective and gives official sanction to that view (Hartsock, 1985, chaps. 9, 10).

Carol Gilligan (1982) lends support to the idea that women are sensitive to the interdependence among people. According to Gilligan, women express a "different voice" in their moral development. Their distinct mode of reasoning derives from their different perception of human problems, an alternative that is not exclusive to women but articulated most clearly by them. This system of morality is grounded in the care of, and sensitivity to, others rather than through a system of abstract rules through which human experiences are evaluated. In the female voice, morality is defined as an ethic of responsibility and is derived from first considering the individual from his or her perspective, judging the situation as "a problem of inclusion rather than one of balancing claims" (p. 160). In the male voice, the first principle is to consider the rule (or law) and then fit the individual problem to that standard. It reflects a sense of separation rather than connection, an "ethic of rights [that] is a manifestation of equal respect, balancing the claims of other and self" (pp. 164-165).

For Diamond and Hartsock (1981), " 'women's work' is characterized by concrete involvement with the necessities of life rather than abstraction from them," which leads to "profoundly different social understandings." For women, "everyday life is more valued, and a sense of connectedness and continuity with other persons and the natural world is central" (p. 718).

Political life, which has been constructed on the basis of the male experience, is perceived differently by the relational, female voice. The distinctions between public and private arenas are not clear. The particular takes precedence over the universal, and rationality is tempered by intuition and feelings. Women engage in politics "as people rooted in networks and communities. . . . Political life *is* community life; politics is attending to the quality of life in households, communities, and workplaces" (Ackelsberg, 1988, pp. 306, 308).

The relational model begins from a point of connection—particular individuals facing life's dilemmas—and recognizes how abstract theories have real, practical consequences. It recognizes and rejects the imposition of a single norm against which all experience should be judged. This leaves room for challenging rules or basing decisions on factors other than strict efficiency or (supposed) neutrality. Intuition or feelings take on legitimacy under such a model. A female morality, which traditionally was confined to the private realm, transforms public life.

If women have experienced life differently, then it should seem obvious that as officeholders they will bring these differences to bear in governance. Within the local politics of the communities studied, the relational approach posed a different set of questions when formulating public policy. Recognizing the impact that laws have on individual lives, and placing a value on such an impact, is reflected in the comments of women councilmembers. Councilwomen hold different attitudes toward alternative concepts of housing; they tend to assess the impact of housing in different terms than do men; and women, in their efforts on behalf of seniors, have focused much of their attention on housing. It must be stressed that in each of these areas, it was not a case of all men thinking one way and the women another, but, clearly, there was an emerging pattern of difference that warranted analysis.

A striking difference between women and men councilmembers was their attitudes toward housing other than single-family residences. This was expressed in their attitudes toward three types of alternative housing: permitting homeowners to take in boarders, high rises, and rental housing. The first alternative was suggested as part of the general interview, but the second two were raised by the councilwomen themselves. Suburban communities are zoned so that areas reserved for single-family homes usually do not permit the boarding of nonfamily members or conversions to two-family residences. Under pressure from a weak economy and burdensome property taxes, money from renting out a room or creating a second apartment is often very attractive to certain homeowners, especially seniors, who often have unused bedrooms, or single parents who seek financial assistance to meet mortgage and property tax payments. Are such violations prevalent in any particular community? With a few exceptions, most women and men agree that they exist, but they react differently.

Men are concerned about enforcing the law. They are concerned that taking in a boarder converts a single-family home into a two-family home, which can lead to "overcrowding and too many cars. You lose control of the whole concept that was drawn up in the beginning." Or "They infringe on safety and the law. . . . They're putting themselves and everybody in jeopardy." And "it's not just that housing values go down. You've got a town of parking cars on the street, causing problems that way; they are using our services and not paying for them." Besides the strain on services, there was the concern that "if there were a fire, all hell would break loose if there were violations." More directly, a councilman worries about the "element" that would be brought into town.

In contrast, some councilwomen have argued either to ignore the practice or legalize it, primarily because they are concerned about "people in need" who are probably forced into taking in boarders because they are "hurting" and "having trouble making ends meet." One councilwoman sees boarders as a way of "giving an apartment, some living quarters to somebody." This concern about the personal reasons for the violations was virtually ignored by the councilmen. Men were far more likely to refer to the "greedy" landlords who were milking their properties for extra money and the absolute need to enforce the law. For women, the "rules" played a less significant role than did the impact of the practice on the lives of people.

The commitment to single-family housing inhibits the construction or conversion of units into low- and moderate-income housing. Although there is general agreement that single-family units are preferable, there is some support among women, and virtually none among the men, for other alternatives. Several councilwomen suggested the possibilities of zoning for buildings higher than four stories, which in suburban terms is the equivalent of treason. Communities that have embraced high rises are held up as examples of what not to become. Yet there were suggestions by women that "[if] we sold the land to a developer and gave him enough density, he could build [some housing for a nonprofit group] that would certainly be comparable to affordable housing, . . . but that is a 'no-no' in this town." Or they have expressed a willingness to tolerate "some apartments" for residents who no longer want to maintain single-family homes. In addition, there was some support for more rentals, including those in high rises, where low-income people are more easily mixed with moderate- and upper-income people and "nobody knows what your income is." Although this kind of support was scattered, and perhaps not uppermost in their minds, they cumulatively stand in contrast to the men, who rejected liberalization of zoning rules or alternative kinds of housing.

The tendency of men to reduce policy decisions to a cost-benefit analysis also arose on occasion. They ruminated about the excess costs to the locality from the illegal residents; they talked about ingenious ways their towns might be able to get money back for building *Mt. Laurel* units; they insisted that their strength on the council, on housing or any other matters, was their "management" ability, which brought rationality to council considerations, serving as an in-house analyst. With two exceptions, the men mentioned their business background and how they brought that experience to their work on the council. Only two women referred to their business background; the rest emphasized other aspects of their experience, even though most of them had been or were employed by businesses, including holding positions with management responsibilities, working in technical areas, and organizing information.

There was also a difference in how women and men evaluated their town's *Mt. Laurel* experience. Men spoke in terms of "units," arguing with "the way they arrived at the numbers," or that the town deserved to receive "more credit for our numbers." A councilman judged his community's affordable housing program as "successful in the numbers that have been built." On the other hand, women were concerned that "the people who needed it didn't get it." Women also expressed concern that the purchase of apartments by many young, recent college graduates who were unlikely to remain in the lower-income brackets, had "perverted what *Mt. Laurel* was all about" and that the results of *Mt. Laurel* had filled the pockets of the developers "rather than those who need housing."

There are some councils where differences produce tension between men and women. Men criticize women for listening to the last sad story and "approach things on a more emotional personalized level." Some women resent that often men "posture" and are concerned that men like "to play the game." Such observations evidence the contrast between the liberal and relational approaches. For women, "need" is a more important value than "management," a view not expressed by most councilmen.

Particularly sensitive to the needs of seniors, councilwomen have been notable in their leadership in this area. In every community, if one of the women is middle-aged (and women on councils are older, primarily because they often do not run for office until after their children are older), often, she has developed a relationship with the senior constituency in that town. This manifests itself in a wide variety of programs, including day care, recreation, and transportation, and in many places it has led to building senior citizens residences, making efforts toward future construction, or working with seniors to alleviate the pressures of paying property taxes while living on fixed incomes. This interest on the part of women resonates perfectly with the prevailing suburban attitudes about *Mt. Laurel*. Municipalities are able to fulfill some of their affordable housing obligation while serving a constituency clearly in need.[6] The senior residents of affordable housing units are often drawn from the nearby areas, and in many places, local residents are given first preference, a practice that will undoubtedly have to stop with the 1993 New Jersey Supreme Court decision that held such a policy unconstitutional. In four towns there was at least one woman who thought of seniors as "her constituency." They express a particular pride in the work they have done for seniors. One councilwoman explained that having "just recently [gone] through the years with my mother . . . which so many of my friends are facing," has made her "aware of the kinds of services . . . people need." Some men have certainly been supportive of these efforts, but in the surveyed towns it is the women who have provided the leadership for most of the projects.

Thus there are some significant differences between women and men in forging housing policy. On a broader note, one councilwoman defined all local issues as "women's issues." As she noted,

> It is the women who are driving in and out of the driveways of the supermarket, dropping the kids off at school, picking them up at school, doing all those things. The women are the ones driving around the community and probably have a keener sense of the planning of a community than most of the men do. They have a sense of the environment that they have been more involved in generally than some of the men have. Even if they're off at work, I'd be willing to bet that they are the ones, who on their way home from work, stop off at the supermarket, not their husbands, picking up the cleaning, picking up the sitter.

And on this basis, she fought to have more women appointed to the local zoning board, and she rebuffed arguments that they needed to fill the opening with an architect or an engineer, arguing that a woman's experience would bring a different perspective to their deliberations.

■ Women in Politics

Relational politics is not consistently evident in the decision making of councilwomen on affordable housing, but it does have a strong presence. Women are more likely to express their displeasure with certain ordinances and their effect on individual lives. Combining the intuitive and the particular puts enormous pressure on the rationality/efficiency norm that is the foundation of low-tax/high-value decision making prevalent at the local level. Women, not all and not always in the same way, have asked and acted from a different viewpoint than do men in making affordable housing decisions.

For women, the sense of community does not extend beyond their municipality's borders. When they perceive that their (narrowly defined) community is threatened, they and their male colleagues engage in liberal politics. When they face inward to those who reside within their localities, then relational politics is evidenced. However, the persistence of liberal politics in the first instance places an enormous obstacle in the path of developing a model of representation that might transform the exclusionary nature of communities that lie outside of central cities. Thus even to the extent that women in local office bring an alternative set of questions or perceptions to housing, they do so severely constrained by the imperatives of liberal thought. They resist anything but the most simple regional accommodations, sharing ambulance equipment, for example. They quite literally compete with other municipalities in terms of service quality, such as "Are their streets plowed quicker and/or better than ours?" "Is their downtown nicer than ours?" "Is their tax rate lower than ours?" "Do they look like what we want to look like?" Moreover, they practice this kind of thinking with an ideal in mind that conceives of public services—good schools, clean streets, safe neighborhoods, Little League—within a quiet, green, single-family setting. This leaves little room to think about the responsibilities of governance as extending beyond their borders. It conceptualizes towns as competitors, not as interdependent communities that need to work toward some common set of goals.

Mt. Laurel is a revolutionary idea that challenged the very soul of this policy, and for the most part, suburban councilwomen and councilwomen responded alike. But not exactly alike. It is true that *Mt. Laurel* as a prod to provide affordable housing in the suburbs to poor inner-city residents has not been realized. However, it is also evident that there have been significant changes in their land use policies and the housing policies that have been adopted. Women have played a critical role in that development. Both women and men have articulated general support for the idea of affordable housing, but their interpretation of the concept is different. Women speak of the pain that individuals experience. Having had different experiences than councilmen, it is not surprising that they have become leading advocates for programs for seniors, that they have been willing to consider other alternatives to the single-family community, that they are concerned that managers of those units built for low-income residents find and keep that population. It may be appropriate to call this a relational kind of politics, one that interprets policies in terms of the impact they have on individuals. The interviews indicate that many men have been critical of this kind of politics, of councilwomen who "respond to feelings and people." Not all women hold these attitudes or act on them, but taken together, clearly, there has been a much greater tendency for women to think about affordable housing in relational terms.

Thus women share certain fundamental attitudes with men, favoring the protection of home rule, private property, and low taxes. However, in spite of this congruence, women's understanding of representation has reshaped the implementation of those commonalities. When focused on intracommunity life, women on councils often see themselves as agents of social fairness rather than as instruments of efficient management. They stress the impact of governmental decisions on individual lives. The "last sad story" is, according to this view, as important to governance as any particular set of supposedly neutral rules. It is not a case of one kind of politics being substituted for the other, of relational politics supplanting liberal politics. Rather, the former has been grafted onto the latter.

Women who assume office on municipal councils find that norms are firmly entrenched and local agendas are mostly fixed. Budgets are presented to them, zoning laws are in place, and priorities are already established. Women react to the established agendas and norms, but the intractability of the status quo constrains the nature of a relationally induced transformation.

Race and class are still powerful engines in the politics of the suburbs. *Mt. Laurel* has done little to dispel that set of attitudes. But *Mt. Laurel* has reshaped the suburban agenda, and as a result, new policies have had to be addressed and old ones rethought. As more women have assumed elected positions on municipal councils, they have responded to *Mt. Laurel*, and they have raised other issues, if not always with a different voice, certainly with a distinctive accent.

NOTES

1. *South Burlington County N.A.A.C.P. et. al. v. Township of Mount Laurel et. al.*, 92 N.J. 158 (1983).

2. The New Jersey Supreme Court held the Fair Housing Act constitutional in *Hills Development Company v. Bernards Township in Somerset County*, 103 N.J. 1 (1986).

3. The research for this study is based on a set of 21 semistructured interviews conducted in five suburban towns limited to one county in New Jersey. Only towns with at least two female councilmembers and that represented a middle-class income level were selected to insure comparability and to expose gender issues. The towns ranged in size from 8,000 to almost 40,000 persons, and the median family income in 1990 ranged from $56,000 to $67,000. Four of the five towns were primarily "White," although, some councilmembers argued that this was an inaccurate characterization. One could be classified as a middle-class racially integrated suburb.

Of those who agreed to be interviewed, there were 10 men and 11 women; 8 Republicans, 11 Democrats, 1 Conservative, and 1 case could not be politically classified. Two communities were nonpartisan, but party affiliation was assigned based on the interview. All the councils consisted of from five to seven members elected at-large. In the county where the interviews were conducted, 19% of the council seats were filled by women.

Interviews were also conducted with state and county officials, and members of advocacy groups.

4. Communities have plans either by applying to and receiving certification from COAH, or they have adopted a plan following suit by a developer in the courts. The towns included in this discussion include both types. Of those that were under the jurisdiction of COAH, not every locality had completed certification.

5. In *In re Township of Warren*, 132 N.J. 1 (1993), the New Jersey Supreme Court declared that giving preference to local residents or workers in affordable housing was unconstitutional.

6. In many communities, senior housing has mitigated *Mt. Laurel* pressures, permitting towns to fulfill some or all of their obligation with senior projects. It was estimated that by the end of 1992, 19% of all *Mt. Laurel* new housing was for seniors (see Fitzpatrick, 1993, p. 4).

REFERENCES

Ackelsberg, M. A. (1984). Women's collaborative activities and city life. In J. A. Flammang (Ed.), *Political women: Current roles in state and local government* (pp. 242-259). Beverly Hills, CA: Sage.

Ackelsberg, M. A. (1988). Communities, resistance, and women's activism: Some implications for a democratic polity. In A. Bookman & S. Morgen (Eds.), *Women and the politics of empowerment* (pp. 297-313). Philadelphia: Temple University Press.

Diamond, I., & Hartsock, N. (1981). Beyond interest in politics: A comment on Virginia Sapiro's "When are interests interesting? The problem of political representation of women." *American Political Science Review, 85,* 717-721.

Disch, L. (1991). Toward a feminist conception of politics. *PS: Political Science and Politics, 24,* 501-504.

Dodson, D. L. (Ed.). (1991). *Gender and policymaking: Studies of women in office*. New Brunswick, NJ: Center for the American Woman and Politics.

Dodson, D., & Carroll, S. J. (1991). *Reshaping the agenda: Women in state legislatures*. New Brunswick, NJ: Center for the American Woman and Politics.

Fitzpatrick, B. (1993, January). Ten years later: Tangible impact of Mount Laurel II. *Housing New Jersey*, pp. 3-5, 18.

Flammang, J. A. (Ed.). (1984). *Political women: Current roles in state and local government*. Beverly Hills, CA: Sage.

Flammang, J. A. (1985). Female officials in the feminist capital: The case of Santa Clara County. *Western Political Quarterly, 38*, 94-118.

Gilligan, C. (1982). *In a different voice*. Cambridge, MA: Harvard University Press.

Hartsock, N. (1985). *Money, sex, and power: Toward a feminist historical materialism*. Boston: Northeastern University Press.

Johnson, M., & Carroll, S. (1978). *Profile of women holding office II*. New Brunswick, NJ: Center for the American Woman and Politics.

Lee, M. (1976). Why few women hold public office. *Political Science Quarterly, 91*, 297-314.

Logan, J. R., & Molotch, H. L. (1987). *Urban fortunes: The political economy of place*. Berkeley: University of California Press.

Merritt, S. (1980). Sex differences in role behavior and policy orientations of suburban officeholders: The effects of women's employment. In D. W. Stewart (Ed.), *Women in local politics* (pp. 115-129). Metuchen, NJ: Scarecrow.

Mezey, S. G. (1978a). Does sex make a difference? A case study of women in politics. *Western Political Quarterly, 31*, 492-501.

Mezey, S. G. (1978b). Support for women's rights policy: An analysis of local politicians. *American Politics Quarterly, 6*, 485-497.

Mezey, S. G. (1978c). Women and representation: The case of Hawaii. *Journal of Politics, 40*, 369-385.

Minow, M. A. (1990). *Making all the difference: Inclusion, exclusion, and American law*. Ithaca, NY: Cornell University Press.

Pitkin, H. F. (1967). *The concept of representation*. Berkeley: University of California Press.

Stanwick, K. A., & Kleeman, K. E. (1983). *Women make a difference*. New Brunswick, NJ: Center for the American Woman and Politics.

Young, I. M. (1990). *Justice and the politics of difference*. Princeton, NJ: Princeton University Press.

8

The Rise and Decline of African American Political Power in Richmond: Race, Class, and Gender

LEWIS A. RANDOLPH
GAYLE T. TATE

The struggle for Black political empowerment in Richmond, Virginia, culminated in the 1977 special councilmanic election of a Black majority city council and Richmond's first Black mayor, Henry Marsh. The interplay of race, class, and gender disrupted the complacency of the White political structure. In seeking to ensure their political and economic viability and eventual economic control of the "new order," the White political structure was forced to adopt new strategies of political survival.

Up until the post-Reconstruction period, Blacks had played a tangential role in the city's politics, most often as mere observers depending on governmental largesse rather than as genuine participants. A White southern oligarchy dominated the prevailing political arena (Key, 1949, p. 19). This White political domination reflected segregated institutional arrangements as well as social superordinate-subordinate relationships between Blacks and Whites, common to southern etiquette, custom, and tradition.

Black empowerment in the post-Civil Rights era, coupled with the advent of Black and White women entering the political arena, destabilized the White political structure and forced changes as biracial and bigender coalitions formed and demanded political inclusion. Black women became major participants in the body politic and eventually were elected to the city council. However, neither the Black political movement nor the women's movement, both of which have touched Richmond's contemporary politics, have satisfactorily addressed Black women's plight. This chapter examines the class cleavages among

Blacks and their impact on gender, which has undermined political progress both in the Black community as a whole and for Black women in particular. The relationship of gender to class and race is demonstrated in the rise and decline of the Black majority city council since the 1970s and in the experiences of one Black woman council member, Willie Dell. Intrinsic to the study is the historical relationship between Blacks and Whites that paced Black political development.

Unlike other urban researchers in their study of cities, we are presenting the dynamics of race, class, and gender and their collective impact on urban and Black politics as major variables altering the political arena and subsequent outcomes of major events. The interplay of these variables offers a comprehensive picture of the evolution of events, the polarization of racial attitudes, prevailing racial cleavages, and the present-day shifting realignments of the Richmond political spectrum. Rothenberg (1988) lends support to this contention when she states, "Using race, class, and gender simultaneously as categories for analyzing reality provides us, at least at this historical moment, with the most adequate and comprehensive understanding of why things occur and whose interests they serve" (p. 37). Moreover, employing these three variables affords the research a rich opportunity to understand the complex political arrangements and linkages across racial, gender, and class demarcations.

This study is an exploratory analysis using a historical framework. The explanatory case study approach was selected as appropriate for a single-case study analysis because of its utility in policy studies, political science, and research studies of cities (Yin, 1989, pp. 13-26). Although critics of case studies are concerned, and appropriately so, with the "biased views influencing the directions of the findings and conclusions" to which this type of examination can fall prey, that concern was mitigated in this study by interviewing a triangular group of established politicians, civic leaders, and grassroots activists to garner a balanced portrait of the city's political life.[1] Although there are limitations inherent in such an approach, we argue that this method allows for the most comprehensive view of the interlocking variables of race, class, and gender in Richmond's diverse political arena.

■ Historical Race Relations in Richmond

The development of White supremacy in Richmond was buttressed by an emerging oligarchy in the early 1800s. This contributed to a superordinate-subordinate relationship between Blacks and Whites as was characteristic of the antebellum era, while simultaneously creating the vehicle for Black political mobilization. V. O. Key, Jr. (1949) stated in his description of Virginia's oligarchical structure that "political power has been closely held by a small group of leaders who, themselves and their predecessors, have subverted democratic institutions and deprived most Virginians of a voice in their government" (p. 19). This politically entrenched planter class of judges, editors, and bank presidents (the Richmond Junto), marked not only conservatism in Richmond's politics and immigrant antipathies but their pseudoaristocratic superiority toward their urban slaves and free Blacks as well (Chesson, 1981, p. 20).

Richmond, the chief manufacturing city in the antebellum South, had a unique system of industrial slavery. By the 1850s, 3,400 hired slaves—more than half of the hired-slave population in Richmond—worked the city's tobacco factories, and others worked in the flour and iron mills, building and construction trades, railroads, and stone quarries, as well as performing domestic services (Rachleff, 1984, pp. 6-7). Earning overtime monies, tobacco workers and iron workers could purchase freedom for themselves and their families.

These semiautonomous urban slaves became vital to the economic and social development of the free Black community by making the transition from slavery to freedom, which for some was by the purchase of real estate (often, however, with the legal title resting in a White person's name) (Jackson, 1942, p. 181). Capital accumulation by both slaves and free Blacks led to the founding of the critical institutions of family, church, and mutual benevolent societies in the segregated free Black community. With freedom, kinship ties, property purchase, and the development of a social network, slaves and free Blacks developed an economic foothold in antebellum Richmond. From this coalition came an emerging Black elite that pressed the White oligarchy for moderate political benefits.

During Reconstruction, the oligarchy prevailing under the guise of the Conservative Party, the military rule of Union troops, and the Republican Party, made pseudoalliances with an emerging Black political elite as they sought to maintain White supremacy in Richmond. This White domination was aided by the political schism in the Black community between the Black elite who were free before the Civil War and those Blacks recently freed. White Republicans invariably exploited this schism between Blacks and betrayed all Black leaders.

Blacks, however, gained suffrage through their alignment with the Republican Party, and many of them were elected to the Virginia General Assembly (Buni, 1967, p. 2).

By 1871, soon after the Conservative Party had regained control of the city from the Republicans, they created the Jackson Ward and gerrymandered its boundaries to include the majority of the city's Black population. Having complete control of the other five wards in the city, the Conservatives concentrated their efforts on minimizing the Republican vote from Jackson Ward. Conservatives, who considered any Black who independently engaged in politics to be radical, strove to handpick moderate Black leaders to serve on city council. Both parties were successful. Even this minor courtship for Black political participation effectively ceased as the Republican Party quickly abandoned its Black constituency, and by 1878, Richmond Blacks were effectively disfranchised (Buni, 1967, p. 1; Morton, 1973, pp. 70-97; Wynes, 1961, p. 145). Even though Blacks continued to serve on the city council until 1898, only "thirty-three black councilmen were elected between 1871 and 1898 from Jackson Ward" (Chesson, 1982, p. 196), and their political effectiveness was muted until 1956.

The Emerging Black Elite

Class and political tensions in the Black community were prevalent but embodied different ideological and pragmatic perspectives. The Black elite was made up of those who were free during the antebellum era. They solidified their economic interests under the rubric of racial solidarity, although their bourgeois respectability pivoted on economic prosperity, light skin color, "southern gentility," and conservatism. In marked contrast, recently freed persons were of darker hue, penniless, lacking social graces, and strident in seeking increased political inclusion and wider distribution of economic resources. Whereas the former group, many of whom were dependent on White patronage, were conservative in their politics and reluctant to press for political change, the latter group, the radical Blacks, advocated disfranchisement of Confederates and confiscation of their property, greater political inclusion, and increased economic viability.

These class cleavages in evidence in the political arena also were intrinsic to Black social life. The Black elite emulated the social mores and culture of White southern society. Their social life, a parody of White pseudoaristocratic trappings, was a closed society barring "outsiders" from participation. "There were social clubs galore and at their entertainment no outsiders were permitted to participate" (Rabinowitz, 1982, pp. 240-241). These distinctive class cleavages of the Black elite and working-class Blacks remain. Its contemporary reconstitution forms the core of today's Black elite who are known as "native Richmonders" (interview with Willie Dell, executive director, Richmond Community Senior Center, August 23, 1991; interview with Saad El-Amin, Richmond attorney, December, 5, 1991; interview with Martin Jewell, vice president of the Crusade for Voters, August 24, 1991; interview with John Moeser, associate professor in the Department of Urban Studies and Planning, Virginia Commonwealth University, August 6, 1993; interview with Roland Turpin, former executive director of the Richmond Housing Authority, August 6, 1991).

The Byrd Machine and the African American Community

According to Derrick A. Bell, Jr. (1980), Blacks are "involuntar[il]y sacrificed" when Whites seek to reestablish fragile relationships (p. 29). This creates a political environment that transcends class cleavages in the White community and stabilizes racial polarization. In this type of political environment, Robert Byrd was elected Governor of Virginia from 1925 to 1929 and U.S. Senator from 1933 to 1966. Hence the establishment of the "Byrd machine" (Buni, 1967, pp. 134-135; Key, 1949, pp. 19-35; Moger, 1968, pp. 330-370; Wilkinson, 1968, pp. 3-61).

Byrd was considered a progressive governor by Virginia's political standards; for example, he advocated a strict law against lynching and refused to engage in "race-baiting" campaign tactics. However, Byrd perpetuated the legacy of White supremacy, although not as overtly as some of his predecessors, by building political power through the suppression of a Black electorate. He sought to control or retard Black suffrage through legal and extralegal measures. He supported the poll tax from 1925 to 1945 that limited the total Black vote, because "the machine owed its existence to a competent management and a restricted electorate" (Buni, 1967, pp. 134-135). Some Blacks who sought to retain their suffrage rights frequently voted with the local arms of the Byrd machine (Key, 1949, p. 33). Other Blacks had discrete associations with the machine that reflected the existing class cleavages and political schisms in Jackson Ward. Still other Blacks had a covetous relationship with the machine. According to Dickinson (1967), a reporter revealed to him that "he saw a list of Negro ministers and real estate men who could be bought, a list which was kept [by the Byrd People] until at least up to the late 1950s" (p. 14).[2] By preventing, controlling, or engaging in duplicitous political participation, the African American electorate was severely circumscribed under the Byrd machine.

Black Political Mobilization

Just as the suppression of Black suffrage retarded Black political development, it also engendered Black political mobilization for social change. Invariably, as Charles Tilly (1978) indicates, the perception of a common enemy facilitates a "defensive mobilization" (p. 72). Black Richmonders moved toward collective goals of participation in the body politic. In pursuit of these objectives, the oligarchy is the "enemy," defined as the "outsider." Black Richmonders, as the collective "we," internalizes the group. "Defensive mobilization" is particularly acute among African American communities within a segregated social structure because it undergirds racial discrimination, sexism, and social isolation (Morrison, 1987, p. 11). This strategy gave rise to organizations such as the Richmond Civic Council and the Crusade for Voters.

The Richmond Civic Council, a coalition of civic associations, fraternal societies, and religious groups, was founded during the 1940s to mobilize voter registration for Black representation on the city council. Even under the ward system, gerrymandering prohibited Black elected councilpersons. The Civic Council contended that increased voter registration would lead to Black representation on the city council and constitute a major political victory to gird mobilization efforts (Moeser, 1982, pp. 32-33). Attorney Oliver Hill was the first African American to be elected to city council since 1898. Although a major political victory, it was diminished by subsequent allegations that the Black clergy that dominated the council was in financial collusion with the White power structure (interview, J. Moeser, August 23, 1991). As a result of these allegations, young Black activists, the new turks, pushed for a new organization that would address the needs of the African American community independent of the city's White leadership.

The Crusade for Voters, founded around 1956, evolved from an earlier ad hoc organization, the Committee to Save Public Schools (Dickinson, 1967, pp. 24-27; Moeser, 1982, p. 34) that had organized to challenge White resistance to the mandates of the two *Brown v. Board of Education* decisions (in 1954 and 1955). Richmond's White community resisted the *Brown* decision by placing a referendum on the statewide ballot to allow local communities to close their schools to circumvent integration. The referendum passed by a majority of voters and rekindled defensive mobilization strategies by Black communities. The Crusade for Voters emerged as an independent political voice for the Black constituency and in so doing, provided an effective challenge to the statewide Byrd machine.

■ The Rise of a Black Majority on City Council

The election of the African American majority to city council can be traced not only to superordinate-subordinate racial relationships and a growing Black population but also to the Richmond Crusade for Voters' mobilization of the African American community. During the 1966 city council election, the crusade elected Henry Marsh, an African American lawyer, and Howard H. Carwine, a White. In 1968, Rev. James G. Carpenter, another White member of the Crusade was elected, thus increasing the crusade's city council membership to three.

The crusade's efforts were augmented by the U.S. Supreme Court in the *Richmond v. U.S. (1974)* decision in which the Court ruled that the city's annexation of land in Chesterfield County, coupled with an at-large council system, was evidence of the city's attempt to dilute the voting strength of Blacks (phone interview with J. Moeser, July 7, 1991). In *Richmond v. U.S. (1977)*, the Court accepted a compromise to change the city's representation from an at-large system to a single-member ward or district system to ensure Black representation on city council (Moeser, 1982, pp. 159-188).

The adoption of a nine-seat single-member district system in 1977 increased Black representation on city council and represented the crusade's biggest achievement; the special councilmanic election called for March 8, 1977 (Moeser, 1982, pp. 174-178). The special election enabled the Crusade for Voters representatives to become an African American majority on city council and to elect the city's first African American mayor, Henry Marsh. Racism in Richmond's politics and annexation strategies galvanized the African American community to politically and legally challenge the Richmond establishment.

After 1977, the African American community maintained a slim majority on the council. Despite this accomplishment, Richmond's Blacks have not achieved political equality. The Black majority on city council sometimes reflected historical divisions and split along class and gender lines. Although class divisions on the city council parallel similar divisions in the Black community, the city council has evinced a pattern of gender tokenism resulting in, generally, one White and one Black female serving as councilpersons at any given time. Female political participation is overtly discouraged, and gender issues are rarely discussed and supported. The gendered divisions in the Black majority on city council and in Richmond's current political environment are traceable to events surrounding the 1982 council election.

The Demise of the African American Majority on the City Council

The collapse of the African American majority on the city council in 1982 can be attributed to White intransigence to Black emphasis on political inclusion, which represented to Whites a radical shift, as well as to racial confrontations over a controversial plan to revitalize downtown, Project One. The class and gender schisms on the Black majority city council and within the Black community eroded the council's solidarity. Members were vulnerable to demands for political change and to the presence of conservative council candidates.

The philosophical and policy differences between Blacks on city council and Richmond's White elites escalated over the 1978 deliberation over the selection of a hotel for the proposed Project One convention center and office complex. The African American majority Council recommended selection of the Mariott Hotel corporation, whereas the White business community backed selection of the Hilton Hotel chain. The conflict escalated when the city council passed an ordinance requiring that all private development, including hotels in the Project One area, be assessed a development fee based on their potential impact on Project One. Moeser (interview, September 11, 1989) asserts that the council passed the ordinance to protect the Mariott chain, believing that other hotels would oversaturate the area and the city would lose money. "On a five to four, black-white split that November, the council denied city permission for the transfer of two parcels of surplus land to the Hilton developers" (Edds, 1987, p. 141).

The Hilton Hotel chain sued the city, and the business community claimed that the city council's actions were racist. The city lost the lawsuit and damages of $5 million dollars were awarded to the Hilton chain. The city council got a vote of "no confidence" in the local newspaper, rekindling the myth that African Americans were incapable of governing. Some Black leaders felt that Mayor Marsh lost stature, particularly in the business community. This inevitably diminished the credibility of the Black majority council and of liberal African American politicians in general.

Some suggested that the loss of the city council's credibility over the development issue triggered the gender-class bias in the Black community that led to the 1982 defeat of Willie Dell, an African American female liberal member of the city council. In essence, she became the sacrificial lamb of the controversy that led to the election of Roy West, a conservative African American, as councilperson and subsequently as mayor of Richmond.

The Election of Willie Dell

Willie Dell, a Richmond "outsider," lost her council seat to Roy West in 1982. She had championed racial, gender, and class issues and had a strong constituency among lower-class and upwardly mobile Blacks. Black community activists believe that Dell's election loss was a major contributing factor to the demise of the Black majority on city council. Dell was a strong supporter of racial solidarity as well as social reform for poor Richmonders. With the exception of Marsh, Dell was the most outspoken leader on city council.

The Black elite, "native Richmonders" agreed with the White business community that her political beliefs were "too radical." Then-Governor L. Douglas Wilder resided in the Third District that Dell represented and publicly disapproved of her leadership style, her radical politics, and her wearing of African attire to council meetings. There is speculation among both Black and White Richmonders that Governor Wilder engineered Dell's defeat (interview, W. Dell, August 7, 1993).

Willie Dell's election hinged on the outcome of a debate with candidate Roy West. She believes that Governor Wilder, who served as moderator, set her up to fail in the debate. According to Dell (interview, August 7, 1993), the crowd was predominantly White, angry, and hostile toward her, and Wilder "drilled" her with "tough" questions—issues not raised by her constituents, such as her accountability to her constituency. For instance, one issue raised was that she did not return their telephone calls. Dell believed that the allegation was incorrect because she had resigned her faculty position to become a full-time elected official for a part-time job that paid her $3,000 less than her former full-time position (interview, W. Dell, 1993). Furthermore, she had office coverage as well as an answering machine and frequently attended community meetings, distributing business cards with both her home and work phone numbers. As the debate progressed, Dell felt that Wilder consistently asked her opponent, West, more general questions, such as, "Roy, as a city council member, how would you deal with the business community in Richmond?" In short, Dell is convinced that Wilder was a biased moderator and had encouraged Roy West to challenge her.

Some observers of Richmond's Black politics accused West and conservative Whites of undermining the Black majority on council as a result of Dell's defeat. The Marsh-led majority on city council was also accused of indirectly contributing to her defeat by failing to secure Dell a satisfactory district when they were redrawn. Richmond was under court order to submit city council district boundary plans every 10 years to the U.S. Justice Department for approval. Under the 1975 boundaries, the population of Dell's district was 90% lower-working-class Blacks. However, when the boundaries were redrawn, Dell's 1982 district was comprised of 69% Black residents and 31% White residents, including more middle-class Blacks and some conservative Whites (see Figures 8.1 and 8.2).

It seems that Dell's district was redrawn to protect Chuck Richardson, a controversial, Black male city councilman whom Mayor Marsh believed was vulnerable to defeat if his district boundaries were not changed. If the population added to Dell's district had been added to Richardson's, then he might have lost instead of her. Dell as a female city councilperson was marginalized in this political exercise. In addition, in the Dell-West contest, voting patterns were typically along class lines. West's support came from the Black middle class; professional groups, such as the teacher's association; and the Black media. He was overwhelmingly supported by the small, older, White, conservative population that had been added to Dell's district (see Table 8.1). "West's margin of victory came decisively in three White precincts in the 69 percent black district. West carried the White precincts by about two to one" (Edds, 1987, p. 143). Dell's support basically came from the poor Blacks in her district. Typically, this area had low voter turnout, which served as an additional contributing factor to Dell's defeat (see Table 8.1). The White population was strongly anti-Marsh. They were unable to defeat Marsh directly but could defeat him indirectly by voting against Dell and for West. Dell lost by 497 votes (Edds, 1987, p. 143).

Dell did not garner Black middle-class support, especially from Black middle-class and upper-class women. One prominent upper-middle-class Black woman expressed her strong dislike for Dell (in one of the

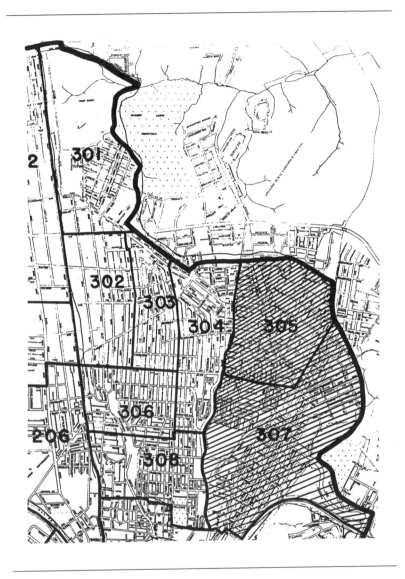

Figure 8.1. Map 1: Map of District Three's Boundaries Prior to 1982

NOTE: Prior to the 1982 election, the boundary lines of Dell's district did not cross Chamberlayne Avenue, the dark line (i.e., street) that separates District Three from District Two. The line-shaded area represents Precincts 305 and 307, both of which were predominately poor and Black and where the voter turnout was generally high. These precincts made up a large part of Dell's political base.

Figure 8.2. Map 2: Map of District Three's Boundaries, 1982

NOTE: For the 1982 election, the boundary lines of Dell's district were extended beyond Chamberlayne Avenue. As shown by the shaded area, Precinct 309 was added to her district and Precincts 307 and 308 were completely redrawn, with both now in the northeastern section of the district. Precincts 307, 308, and 309 were composed primarily of conservative White voters who were not part of Dell's district in 1980. The addition of these new voters to the district, along with the strong upper-income and middle-class anti-Dell vote in Precincts 301 and 302 and low voter turnout in almost all the other precincts, contributed to Dell's defeat and thus shifted the balance of power on the city council.

TABLE 8.1 Richmond City Council District 3: 1982 Election Results, by Precinct

	301	302	303	304	305	306	307	308	309	Total
Dell	316	336	527	656	751	544	78	86	39	3,333
West	344	372	245	190	247	218	571	841	790	3,818
Total number of registered voters	1,328	1,387	1,294	1,525	1,752	1,857	1,450	1,628	1,512	13,773
Total number of voters who voted	660	708	772	846	998	762	649	927	829	7,151

SOURCE: Registrar's Office, City of Richmond, Richmond, Virginia, August 6, 1993.
NOTE: The results of the election returns clearly show that Dell lost by 485 votes. Moreover, the data also reveal that low voter turnouts in Precincts 301, 303, 304, 305, and 306 prevented her from being reelected. In fact, if 500 more voters had voted in either 305 or 306, Dell would have retained her seat on the city council. Thus some of the blame for defeat can actually be attributed to a lack of support from her political base.

author's presence during an interview), declaring that Dell lost her election because Dell was an "outsider," and that the Black women she knew in Richmond did not think that Dell talked or dressed like a lady or a "proper" preacher's wife. On another occasion, during a meet-the-candidate forum before an audience of middle-class and upper-class Blacks, Dell was "hissed" and "booed" and advised to leave before the meeting deteriorated (interview, W. Dell, August 7, 1993; phone interview, S. El-Amin, January 14, 1992).

Dell was unable to rally a constituency divided along racial and class lines. Even her status as a female candidate could not overcome the class expectations of middle- and upper-class Black and White women voters in the district. In some instances, attacks on Dell were gender-biased from both the Black and White communities. Dell was accused of serving on council because she "had an unhappy home life" or was "looking for a man" (interview, W. Dell, August 7, 1993). Yet neither West, Marsh, nor other elected councilpersons, were accused of "looking for a woman" as their reasons for seeking an elected position. Moreover, the sexist overtones of the Dell-West campaign were so evident that both White and Black observers of Black politics in Richmond could not recall such attacks and slander against a male candidate (interview, W. Dell, August 7, 1993).

A final factor contributing to Dell's defeat as well as the destabilization of the Black majority on council was the political rivalry between

Wilder and Marsh. Dell served as a political foil against Marsh for Wilder, because she was one of Marsh's most ardent supporters on the city council. The rift between Marsh and Wilder was precipitated by two factors. Marsh failed to attend a political "summit" meeting between the council and Wilder when Wilder was a state senator. In addition, Marsh did not accept Wilder's advice on the Project One controversy. This rift prompted Wilder to surreptitiously support West for council and defeat Willie Dell. His actions indirectly contributed to the destabilization of the Black majority on city council. Wilder denies the allegation, but leaders in both the Black and White communities believe it (interview, W. Dell, August 7, 1993).

Willie Dell's opponents correctly perceived her political vulnerability. Factors such as the redrawn district lines, the dissatisfaction with her performance as a council member by her new constituents, Wilder's support for her opponents, and her close ties with Marsh, played a decisive role in her defeat. Dell may have been targeted because, next to Marsh, she was the most outspoken member of the Black majority council, and she was female. The perception of her as a "domineering" female, her status as an outsider, along with her involvement in the controversy surrounding the Black majority's handling of Project One, made her politically vulnerable.

The gender-biased campaign politics had a tremendous impact on Dell's life. The campaign increased her blood pressure, there were threats against her life, sexual accusations were made against her, and her husband's masculinity was attacked when she was accused of "wearing the pants" in her household. She was accused of being the stereotypical Black "matriarch" or "sapphire" (interview, W. Dell, August 7, 1993). As Paula Giddings (1984) explains, this negative stereotyping is often used as a ploy by Black men to prevent independent Black women from entering politics and once entered, used as a "red herring" to discredit African American women from further political involvement (p. 339). Black women are generally discouraged in politics because of their inability to build a strong Black support base among Black men who are more vocal about their dislike of female candidates. Black women then seek other avenues of political expression. The gender divisions in the Black community solidified with the White elite and muted her political voice.

Blacks perceived that the manipulation of West was orchestrated by conservative Whites in an attempt to undermine the Black voting bloc on the city council and their reform efforts. Willie Dell substantiates this view. According to Dell, a member of Team Progress (a conservative, predominantly White interest group opposed to Marsh and the Black majority on council), informed her after she lost her election that Team Progress considered her political beliefs "too Black" and that as a result of her outspoken views, White conservatives had targeted her for defeat. The White business elite targeted Dell as being "too radical and too Black" (interview, W. Dell, August 7, 1993).

Team Progress supported Melvin Law, a Richmond Black conservative that Dell defeated in the councilmanic election of 1980, rather than West. Once West was elected, however, they did persuade him to run for mayor and defeat Henry Marsh. "At one after-hours session at the Universal Leaf Company headquarters, about ten executives in various Richmond businesses urged West to join with the four whites on council to elect himself [as Mayor]. West gave no commitment" (Edds, 1987, p. 143). Subsequently, however, West voted with the White council members to elect himself mayor, unseating Henry Marsh, and destroying the Black majority voting bloc.

With the election of West (and the loss of Dell) and his subsequent alliance with the White conservative city council faction, the Black voting bloc, on which the Black majority city council rested, was destroyed. No longer able to wield effective voting power as a collective, the city council reform efforts ceased (interview, S. El-Amin, December 5, 1991; interview, M. Jewell, August, 24, 1991; interview with George Lee, Richmond attorney, August 21, 1991; interview, J. Moeser, August 23, 1991; interview with Michael P. Williams, staff writer, *Richmond Times-Dispatch*, August 24, 1991). As a result of West's election, this new Black conservatism sent a clear political signal to the Richmond business community. Marsh was "de-throned" and they could return to conducting business as usual.

Race, class, and gender factors were intrinsic to Dell's defeat and West's eventual election as Mayor of Richmond and overarched the political and economic domination sought by the White elites. The racial polarization of Richmond bolstered the decision by some White business leaders to support Black conservative candidates against Dell and illustrates their determination to destabilize the Black majority city council and to restore their political and economic control over the city council's policies. Similarly, gender and class divisions in the Black community unconsciously served the interests of White conservatism by targeting an outspoken female, Willie Dell, for defeat. This served to destroy the first opportunity that African Americans, and in particular a female elected official, had to determine reformative public policy in Richmond.

■ Political Implications of the 1982 City Council Election

The primary goal of this chapter was to examine the role of race, class, and gender in the rise and decline of the African American majority on Richmond's city council. The research revealed that race contributed to the election of the Black majority on council, by mobilizing African Americans to vote and subsequently to defeat White reactionary politicians. Race was also used indirectly and negatively by Whites who exploited the existing class and ideological cleavages that divided Richmond's Black community. Gender was used to increase those cleavages by exploiting the candidacy of Willie Dell. Thus, when gender is added to such divisive issues as class and race, it is apparent that all three factors contributed to the demise of the Black majority council. Important, however, is that the political development of female council representatives was sacrificed to overcome race and class cleavages.

Short- and Long-Term Effects

Rightly or wrongly, Dell's defeat signaled to the business elite of Richmond the end of confrontations between city council and the downtown business community. The conservative White council members, with the assistance of a conservative Black, Roy West, hastened this change. The hiring of new public officials solidified the new conservatism. West was now in the position to end the dispute over the Project One hotel agreement, opposed by Marsh and the remaining Black council members. He agreed that the city would pay $5 million dollars to the Hilton Hotel chain. Not unexpectedly, the new majority on council has cooperated with the private sector in economic revitalization projects (Sauder, 1991).

A major result of the 1982 election is that it eliminated the effective voice of African American women in Richmond politics for a decade. Dell has been reluctant to reenter politics after her defeat to West. Claudette McDaniel, the other female member of council during the Marsh years, was defeated in 1989 but did not represent the same type of constituency and issues that Willie Dell had represented. An African American woman would not join city council again until 1993 when the

Rev. Gwendolyn C. Hedgepeth was elected (interview with Hazel Dab-
ney, political reporter, *Richmond Press*, August 6, 1993).

The long-term policy implication of the 1982 election is that Dell's
defeat allowed the business and corporate elites to regain control of the
city's development policy, noticeably in the creation of a quasi-public
corporation, Richmond Renaissance. Although initially conceived by
Henry Marsh, Richmond Renaissance is controlled by a consortium of
corporate leaders and conservative African American elites.

The 1982 election destroyed the Black voting bloc established under
Marsh and ushered in a new Black conservatism. The "set-aside" program for minority contractors, initiated by Marsh, and implemented
during the West administration, was ruled unconstitutional by the U.S.
Supreme Court in *Croson v. Richmond* (Drake & Holsworth, 1991). The
Croson decision further restricted the council's ability to engage in
redistributive policies for Richmond. According to some public officials, Black professionals clearly obtained more contracts from the city,
but they usually were awarded to West's backers.

■ Conclusion

The African American struggle for political and social reform in
Richmond has been circumscribed by White political and economic
domination. This has led to a collusion of some Black leaders with the
White elite and a secondary status afforded to the political voice of
Black women. The superordinate-subordinate relationships still exist,
although they have been given a new patina. Clearly, the White elite
was unwilling to share power with the Black majority city council and
sought their own political salvation in the destruction of the city council's reform efforts. This issue should be explored further.

The gender bias in the Black community is a manifestation of existing class cleavages. It is exposed in the struggle of the community to
maintain its pseudotrappings of gentility that assuaged the void created
when the community was denied liberal reforms in the political arena.
No matter what type of development projects offered by Black politicians, if the economic impact of those projects was not totally controlled
by the White elite with Blacks receiving token economic benefits, any
project would have been treated akin to other reform efforts and would
have been denied. Willie Dell's defeat should be viewed in the context
of liberal reforms denied to the Black community.

In various ways, the Richmond case points to the suppression of the Black women's voice. Race, class, and gender were intertwining factors in her defeat. Most crippling was the double standard of measurement—a sexist barometer—that was applied to Dell when her personal and political motives were circumscribed and questioned. Although all three factors were paramount in this study, the exploitation of gender in the electoral arena as a means to divert attention from class and race division should be explored further.

NOTES

1. Interviews were further substantiated by primary and secondary source materials to separate facts from opinions and biased interpretations of actual events. In addition to the structured and semistructured interviews, the primary data of institutional statistics and documents, doctoral dissertations, master's theses, and unpublished papers were gathered over a period of 8 weeks and three field trips to Richmond. Secondary sources included pamphlets, newsletters, monographs, and newspapers, as well as interviews with media experts on local Black politics and urban scholars.

2. Dickinson's (1967) master's thesis researched the Crusade for Voters, the group that spearheaded the Black political majority city council.

REFERENCES

Bell, D. A., Jr. (1980). *Race, racism and American law*. Boston: Little, Brown.

Buni, A. (1967). *The Negro in Virginia politics: 1902-1965*. Charlottesville: University of Virginia Press.

Chesson, M. B. (1981). *Richmond after the war, 1865-1890*. Richmond: Virginia State Library.

Chesson, M. B. (1982). Richmond's Black councilmen, 1871-1896. In H. N. Rabinowitz (Ed.), *Southern Black leaders of the reconstruction era: Blacks in the new world* (pp. 191-221). Urbana: University of Illinois Press.

Dickinson, A. J. (1967). *Myth and manipulation: The story of the Crusade for Voters in Richmond, Virginia*. Unpublished Scholar of the House Paper, Yale University.

Drake, A., & Holsworth, R. D. (1991). Electoral politics, affirmative action, and the Supreme Court: The case of *Richmond v. Croson*. *National Political Science Review, 2*, 65-91.

Edds, M. (1987). *What really happened when civil rights came to southern politics*. Bethesda, MD: Alder & Alder.

Giddings, P. (1984). *When and where I enter: The impact of Black women on race and sex in America*. New York: Bantam.

Jackson, L. P. (1942). *Free Negro labor and property holding in Virginia 1830-1860*. New York: D. Appleton-Century.

Key, V. O., Jr. (1949). *Southern politics: In state and nation*. New York: Knopf.

Moeser, J. V. (1982). *The politics of annexation: Oligarchic power in a southern city.* Cambridge, MA: Schenkman.

Moger, A. W. (1968). *Virginian: Bourbonism to Byrd 1875-1925.* Charlottesville: University Press of Virginia.

Morrison, M.K.C. (1987). *Black political mobilization: Leadership, power, & mass behavior.* Albany: State University of New York Press.

Morton, R. L. (1973). *The Negro in Virginia politics: 1865-1902.* Spartanburg, SC: University of Virginia Press.

Rabinowitz, H. N. (Ed). (1982). *Southern Black leaders of the reconstruction era: Blacks in the new world.* Urbana: University of Illinois Press.

Rachleff, P. J. (1984). *Black labor in the South: Richmond, Virginia, 1865-1890.* Philadelphia, PA: Temple University Press.

Rothenberg, P. (1988). Integrating the study of race, gender, and class: Some preliminary observations. *Feminist Teacher, 3,* 37-42.

Sauder, R. (1991, December 20). Marsh quits council early in surprise. *Richmond Times-Dispatch,* pp. 1-3.

Tilly, C. (1978). *From mobilization to revolution.* Reading, PA: Addison-Wesley.

Wilkinson, J. H. (1968). *Harry Byrd and the changing face of Virginia politics 1945-1966.* Charlottesville: University of Virginia Press.

Wynes, C. E. (1961). *Race relations in Virginia: 1870-1902.* Charlottesville: University of Virginia Press.

Yin, R. K. (1989). *Case study research: Design and methods.* Newbury Park, CA: Sage.

9

Electing Women to Local Office

SUSAN A. MacMANUS
CHARLES S. BULLOCK III

In two decades, women have gone from political obscurity to prominence in many communities. Although women have made great strides in winning executive and legislative positions in cities, counties, and school districts, the "depth" of our knowledge about the "gender transformation" at the grassroots level is much shallower than it should be, especially compared with studies of women in state legislatures and Congress.

Perhaps the sheer number of local governments relative to 50 state legislatures or one U.S. Congress has dampened researchers' enthusiasm for in-depth studies of women in local office. The U.S. Bureau of the Census identifies 3,042 counties, 19,200 municipalities, 16,921 townships and towns, and 14,721 school districts. These thousands of *local* offices have been the point of entry for many women into the political arena.

■ The Study

Little is known about the commonalties and differences in the electoral success stories of women holding local offices. Are some types of positions "easier to win" than others? If so, why? Do certain electoral system structural features (at-large elections, runoffs, staggered terms, election timing, etc.) affect women's local political success? Are gender-based role expectations, explanations, and policy preferences the same as men's? In this chapter, we examine these questions using research on women elected to city, county, and school district governments. We conclude with a research agenda.

Data Sources

Most statistics on women holding local office are based on *membership* surveys conducted by professional associations, such as the International City/County Management Association (ICMA), the National Association of Counties, the National League of Cities, the National Conference of Mayors, the National School Boards Association, the Joint Center for Political and Economic Studies (JCPES; Black elected officials), the National Association of Latino Elected and Appointed Officials (NALEO), the National Women's Political Caucus (NWPC), and the Center for the American Woman and Politics (CAWP). Occasionally, urban scholars conduct their own studies, but these efforts are limited by cost and access. However, these more scholarly investigations tend to go beyond numerical trends to consider factors promoting women's presence in local offices. They also are more likely to investigate gender-based, role and policy preference differences of elected officials.

■ Women Elected to Office in Municipalities

Practitioner associations and scholars alike have focused more on elected officials in cities (mayors, city council members) than in counties or on school boards.

Women Mayors

Women mayors are no longer news as "firsts" or "path-breakers." The NWPC (1993, p. 7) found that women mayors in cities with populations of 30,000 or more increased from 12 in 1973 to 175 in 1993, or from 1.6% to 18.0%. When smaller jurisdictions are included, the percentage of women mayors drops. The ICMA 1991 Form of Government Survey observed that 12.8% of the mayors in cities of 2,500 and more (and cities of less than 2,500 with city managers) were females (MacManus & Bullock, 1993).

Women have won mayorships in big cities, which challenges notions that the "bigger the prize," the less likely women are to win. As of January 1992, 19 of the nation's 100 *largest* cities had women mayors (CAWP, 1992, p. 2). In the past, women mayors have served in such megalopolises as Chicago, Houston, Dallas, San Francisco, and San

Antonio. Some mayoral contests are multiwomen affairs, with nearly one fourth of Florida's female mayors having female opposition in their most recent contest (MacManus, 1992, p. 162).

Who Wins and Where?

The 1991 ICMA survey data (the most inclusive data base available—4,967 cities) found women most likely to be elected in cities with populations from 250,000 to 499,999 (31.3%) and from 100,000 to 249,999 (20%) (MacManus & Bullock, 1993).[1] For cities under 5,000, the percentage drops to just over 10%. In terms of region, women mayors are most prevalent in New England and along the Pacific Coast. Almost one fifth of the mayors in these regions were women, whereas, at the other extreme, only 6% of the mayors in the East South Central region (Alabama, Kentucky, Mississippi, and Tennessee) were female. Past studies have found that the more traditional culture in the Deep South discourages the election of women (Carver, 1979; Hill, 1981), although women have gained ground even there (MacManus & Bullock, 1989).

Women mayors are more educated than their constituents, especially in larger jurisdictions (Carroll, 1985, p. 66). Women politicians, more than men, tend to be drawn from among educators, whereas more men than women come from the legal profession. The ICMA data show that a higher proportion of women than men had master's degrees.

Few female mayors in the 1991 ICMA survey are racial or ethnic minorities. Of the 630 cities with women mayors identifying their race/ethnicity, 605 were White, 14 were Black, 9 were Hispanic, and 1 each was a Native American and an Asian. This distribution was almost identical to that for male mayors. Data collected annually by the JCPES since the early 1970s showed that Black female mayors increased from 9 in 1975 to 75 in 1992. Black female mayors tend to head smaller, poorer towns in the rural South where large concentrations of Black voters live (Jennings, 1991). In 1988, only 5 Black women governed cities with populations over 30,000. The median population of cities electing Black women mayors was 2,218; the median per capita income, $5,862.

Few studies have examined the age profiles of women mayors. Johnson and Carroll (1978) found that in 1977, the median age of 285 women mayors was 52, with 60% 50 years of age or older. Older women were more likely to head larger cities and were, on average, older than council members. However, a recent study of Florida cities found that a higher proportion of female mayors (50.1%) than council members (34.4%) were younger than 46. The same study found that the age of entry into politics was slightly lower among female mayors than among female council members. Younger women today are more likely to have higher political ambitions than older women, which may explain these patterns (MacManus, 1991; MacManus & Bullock, 1992).

Surprisingly, few studies have focused on the political party affiliations of female mayors. Johnson and Carroll (1978, p. 21A) showed that in 1977, 50% were Democrats, 33% Republicans, and 17% Independents—results echoed in the 1990 survey of Florida mayors (MacManus, 1992, p. 162). Democratic mayors are more likely to head larger, more heterogeneous central cities; Republicans, smaller, more suburban communities, reflecting the partisan makeup differentials commonly associated with these two types of municipalities.

Among women mayors in 1977, 20% labeled themselves liberal or very liberal, 43% as middle-of-the-road, and 37% as conservative or very conservative (Johnson & Carroll, 1978, p. 34A). A propensity toward moderation and "political pragmatism" characterized women mayors in Florida in 1990. When a fourth category ("changes with the issue at hand") was added to the three usual ideological classes, nearly half (46.7%) classified their ideology as "changing with the issue at hand." Only 6.7% labeled themselves liberal; 33.3% were moderate, and 12.5% were conservative. This may reflect the inclination of younger politicians to be social liberals but fiscal conservatives, or it may show that chief executives realize the need to work out compromises to win approval for their proposals (MacManus, 1992).

Full or Limited Legal Authority?

Mayoral powers (budget, appointment, veto, succession) vary considerably across American cities and forms of government (strong mayor-council, weak mayor-council, commission, council-manager). The 1991 ICMA survey (MacManus & Bullock, 1993) revealed that women more often head cities with town meeting and council-manager forms of government (20.5% and 15.3%) than mayor-council structures (10.0%). There is no breakdown between strong- and weak-mayor-council structures. Women mayors serve on city councils in a slightly larger percentage (67.3%) than do male mayors (60.2%). Mayoral service on city council is a common feature of commission and council-manager forms of government. Just over 15% of all mayors (men and women) serve full-time, usually in large jurisdictions with strong-mayor forms of government. More women than men mayors are chosen by the council rather than directly by the voters (26.3% compared with 20.3%). This phenomenon complements one that is observable on school boards—greater diversity in appointed than elected systems (Robinson & England, 1981). Only 5% of all mayors face term limits, a barrier equally likely to affect men and women chief executives.

Women are not precluded from achieving powerful positions, when measured in terms of mayoral voting and veto powers. More women than men mayors can vote on *all* issues (59.8% vs. 48.9%). More male (40%) than female (28.7%) mayors are restricted to voting only to break a tie. Male mayors are slightly more likely to have a veto (31.9%) than are women (26.3%), probably because more men than women head large strong-mayor council governments. In general, however, there are few sizable gender-based differences in the powers wielded by mayors.

Political and Electoral Factors Affecting Election

A successful "first" woman mayor was seen as smoothing the path for women city council candidates (MacManus, 1981). But more recent studies posit that women mayors are more likely in cities where the proportion of women serving on the city council is higher, because women mayoral candidates often emerge from the ranks of council members (Bledsoe & Herring, 1990). The 1991 ICMA survey found that across every population category, a higher proportion of women serve on councils where there are women mayors (MacManus & Bullock, 1993).

The perception that one needs prior service in an elective capacity is increasingly more "fiction than fact." Among women mayors in 1977, 30% had held another elective post prior to being elected mayor (36% of these had held a city council seat) (Johnson & Carroll, 1978, p. 22A). Among a sample of women mayors in Florida in 1990, only 6.3% had previously held elected positions. Participation in high-profile community organizations and activities and political party organizations may be better "training grounds" today than lesser elective posts. The Johnson-Carroll study in the late 1970s showed that 28% of the women mayors held at least one political party office (mostly at the city and county level). Their study also showed that women were more active in a wider variety of organizations than are men. Among women mayors, 38%

belonged to a political organization, 30% to a business and professional organization, 27% to service organizations, and 22% to youth and school service organizations. Participation in women's networks was extensive and involved two thirds of the mayors.

Black women mayors face numerous obstacles. Among these are

(1) fewer opportunities to acquire the political training, experience, skills, knowledge and resources necessary to move into the political arena; (2) discrimination due to racism and sexism—the "double whammy"; (3) the failure of political parties to recruit them to run in viable contests; (4) the difficulty of raising money to run effective campaigns; and (5) the power of incumbency in races where they are most likely challengers rather than incumbents. (Jennings, 1991, p. 73)

These common perceptions need further empirical testing.

Roles and Policy Preferences

There is growing evidence that "women and men [mayors] see their political environments—the constituency and the problems their constituencies face—in the same way" (Rinehart, 1991, p. 96). After contrasting the views of five pairs of male and female mayors who had served the same city, Rinehart (1991) concludes:

It is not surprising that men and women who have been active for, campaigned in and led the same city, would be in close agreement. . . . These are people who knew their cities well enough to win the confidence of the electorate. (p. 96)

When differences were perceived, four of the five women mayors "believed that women bring an added, or extra dimension to politics" and that "women were substantively different from men" (p. 101).

Black women mayors "see themselves as using different leadership styles and having different ways of solving problems than their towns' previous male mayors," are more committed to addressing problems of the economically disadvantaged, and see themselves as more likely to use team approaches to solve problems (Jennings, 1991, p. 73). These women (like their White counterparts) "said that the people of the community felt very comfortable talking to them, they were accessible to the public, and the constituents could talk more freely to them than to previous [male] mayors" (Jennings, 1991, p. 77). They came to office following involvement as school teachers, community activists, and church leaders.

Women City Council Members

In one of the first studies of female council representation, Karnig and Walter (1976) found that women held 9.7% of the council seats in cities with populations 25,000 and more. By 1978, women filled 12.8% of the council seats in cities with populations of 25,000 and more and with at least 10% Black populations (Karnig & Welch, 1979). In the mid-1980s, the figure had increased to 18% in cities with populations of at least 25,000 (Bullock & MacManus, 1991). By that time, almost three fourths of these cities had women on their councils, up from 56% in 1978. The 1991 ICMA survey (cities of all sizes) showed that 70% of the respondent cities had at least one woman on council; the larger the population, the higher the percentage. The proportion of female council members in 1991 was 18.7% (MacManus & Bullock, 1993), up from 12.8% in the 1981 ICMA survey.

Who Wins and Where?

One of the long-standing questions has been whether women are more successful in winning posts in smaller or larger municipalities. Those who believe females win more often in smaller jurisdictions base their opinions on research suggesting that "women's representation is negatively related to the prestige of the office" (Karnig & Walter, 1976, p. 609), thereby making it easier for women to win in smaller jurisdictions where such posts are less prestigious and competition less keen.

Others have asserted just the opposite—that women win more council seats in larger jurisdictions:

> Larger communities tend to be more cosmopolitan and less traditional, thus perhaps more open to women officeholders. In larger towns, there are more likely to be groups such as the League of Women Voters, the National Women's Political Caucus, and the American Association of University Women that would give support to women candidates. (Darcy, Welch, & Clark, 1987, p. 40)

Yet another view is that size of jurisdiction makes little difference in the competitiveness of the council seats (MacManus & Bullock, 1992).

Data from the 1991 ICMA survey show women more likely to serve in cities with populations of 25,000 or more where women hold at least 22% of the council seats, whereas the mean for the smaller jurisdictions is 19% or less (MacManus & Bullock, 1993). This pattern is in keeping with the proposition that women are more likely to be elected in urban areas, because of less sexism in the electorate and/or the greater likelihood that politically oriented women's groups will be larger and more active in bigger cities.

Women councilors are best represented in jurisdictions governed by representative town meetings (24.2%), followed by council-manager governments (20.2%), town meetings (18.4%), and mayor-council jurisdictions (17.4%). But if female representation on councils in mayor-council cities over 500,000 (23.7%) is separated from that in mayor-council cities under 500,000 (17.3%), the picture is quite different. Mayor-council governments in very large cities yield higher female council representation levels.

There are far fewer studies of the backgrounds of councilwomen than of female mayors, with the most extensive studies being collected in nationwide undertakings by Johnson and Carroll in 1977 and Welch and Bledsoe in 1982. These studies discovered some of the same patterns that characterize women mayors with regard to education, race, age, party affiliation, and ideology. Women city council members are more educated than their constituents. Johnson and Carroll found that 63% of women council members had at least some college (the figure was identical for women mayors). Women in smaller jurisdictions had less formal education than did women councilors in larger jurisdictions.

Women city councilors are younger than women mayors, according to the Johnson-Carroll study (1978). About one fourth of the women council members in 1977 were aged 39 or younger, compared with 15% of the women mayors. A 1990 study of a sample of Florida women mayors and council members reported that 34% of the female council members were younger than age 46 (MacManus, 1992).

Women councilors virtually mirror women mayors in their political party affiliations. In 1977, 51% of the female council members labeled themselves Democrats, 33% Republicans, and 15% Independent or Other (Johnson & Carroll, 1978). The 1982 study found that nearly equal proportions of both men and women councilors identified with each political party: 30% were Republicans, 45% Democrats, and the rest Independents (Welch & Bledsoe, 1988). The 1990 Florida study found more Republican councilwomen (51.7%) than Democratic (41.3%), probably due to more suburban communities among the Florida munici-palities sampled (MacManus, 1992).

The ideological breakdown of women councilors varies depending on the time of the study and the database. In 1977, 27% of a national sample classified themselves as liberal or very liberal, 35% as middle-of-the-road, and 38% as conservative or very conservative (Johnson & Carroll, 1978). In 1982, 42% of the women councilors in a national survey of cities with populations between 50,000 and 1 million (Welch & Bledsoe, 1988) saw themselves as liberal, 22% as middle-of-the-road, and 36% as conservative. In contrast, among their male counterparts, 26% said they were liberal, 19% middle-of-the-road, and 55% conser-vative. The 1990 Florida survey, which added the category "changes with issues at hand," showed that 39.1% classified their ideology as shifting with the issue at hand, 39.1% as moderate, 23.4% as conservative, and only 9.4% as liberal (MacManus, 1992).

Much better data are available showing changes in the racial and ethnic composition of female city council members. The 1977 study of cities of all sizes found that only 2% of women city council members were Black and less than 5% Hispanics (Johnson & Carroll, 1978). A study based on 1988 data (collected by survey from 314 U.S. cities having populations of 50,000 or more with Black or Hispanic popula-tions of at least 5%) found that Black women held about 5% of the council posts, Hispanic women 2%, and Anglo women 18% (Welch & Herrick, 1992, p. 157).

The JCPES identified 763 Black female city council members in 1992. A 1993 study by NALEO showed a marked increase in the number of female Hispanics serving in city government between 1986 and 1993 from 138 to 319.[2] Minority electoral success is greater where a minority group's population is larger, more geographically concentrated, and politically cohesive (MacManus & Bullock, 1993).

Legal Factors Affecting Election

The proportion of women on councils and the structural features that affect the election of women to city office have been extensively researched. There has been speculation that nonpartisan elections de-crease female council representation, because a Republican, higher socioeconomic, White, male bias has been identified among candidates elected in nonpartisan races (Cassel, 1985; Welch & Bledsoe, 1988). Alternatively, in nonpartisan settings, candidates must rely more on networks of friends, fellow workers, and associates than on political parties. Because women participate more in civic activities than do men, they are assumed to have an edge. Because women tend to be active in local party organizations and are equally likely to be recruited by party leaders, nonpartisan elections may have no gender-based impact. The 1991 ICMA survey confirms that nonpartisan elections do not keep women off city councils. In nonpartisan municipalities, women hold 18.9% of the council seats, compared with 18.1% in jurisdictions that elect by party (MacManus & Bullock, 1993).

Other structures associated with the Progressive reform movement, such as at-large elections and staggered terms, also have turned out to be less detrimental for women candidates than initially feared by some women's advocacy groups. No structural feature has been studied more than the method of electing city councilors. Borrowing from the literature examining minority council representation, urban scholars have probed extensively the effect of single-member districts, pure at-large (multimember) formats, mixed systems (where some councilors are elected at-large and others from single-member districts), at-large residency formats (which require that a person live in a specific district but be voted upon citywide), and at-large post formats (where a person runs citywide but for a specific numbered post) on female representation levels.

At-large electoral formats have generally been found to *promote* women's election, albeit only weakly (see Bullock & MacManus, 1991; Darcy et al., 1987; MacManus & Bullock, 1993, for a thorough review of this extensive literature.) The 1991 ICMA survey shows that in municipalities where all council members are nominated and elected at-large, women hold 19.7% of the seats. Where council members are nominated and elected by single-member districts, women are least prevalent, occupying 16.3% of the seats (MacManus & Bullock, 1993). A study based on 1988 data that examined the impact of council election method on women of different races and ethnicities found that "electoral structures have very little impact on the representation of Black, Hispanic, and Anglo women." The researchers conclude that "if anything, women of all ethnic groups fare better with at-large systems, but the differences . . . as in the past, are quite small" (Welch & Herrick, 1992, p. 161).

One common question is whether women council candidates do better if they run for seats on larger councils. Those touting a desirability, or prestige, thesis posit that it is easier to win where there are more council seats because competition is less, leading to the conclusion that the office is less prestigious (articulated in Welch & Karnig, 1979). Skeptics answer that larger councils tend to be in larger cities, "where virtually no office is considered to be lacking in prestige if for no other reason than heavier media exposure" (MacManus & Bullock, 1993, p. 78). They also anticipate greater female representation on larger councils due to the more cosmopolitan nature of the electorate and the presence of more female political support groups.

Most studies that include council size as a variable have found a weak, although positive, relationship to female representation. In the smallest councils (those with fewer than four members) surveyed in 1991, women hold 14.9% of the seats. In the modal category, councils from four to six members, women hold 18.3% of the seats. Women occupy 19.6% of the seats in councils that have between seven and nine members. They are most prevalent on councils that have more than nine members, holding 20.8% of these seats (MacManus & Bullock, 1993, p. 78).

A number of studies have tested the hypothesis that the longer the term, the more desirable and competitive an elective office is presumed to be; therefore, more women will be elected to seats with shorter terms than to those with longer terms. Few studies have found any significant relationship between term length and the proportion of women serving on city council. In 1991, the ICMA found that the proportion of women serving on councils in cities where the term is from 1 to 4 years (the overwhelming majority) falls within a narrow range—from 16.7% to 20%. Only in jurisdictions with exceptionally long council terms (5 years or longer) are there significantly fewer women on the council (11.1%) (MacManus & Bullock, 1993).

Is it better for a female to run in a city where council elections are staggered or in a place that elects all its council members at the same time (simultaneous or coterminous elections)? Proponents of staggered terms maintain that they make it easier for challengers (many of whom are women) to gain name recognition and to beat incumbents, especially in smaller jurisdictions. Supporters of simultaneous terms argue that such a configuration makes slating and group coalition building easier and also reduces campaign costs if group advertising is used. Few studies have found that term structure makes any substantial difference in female representation levels (Bullock & MacManus, 1991; MacManus & Bullock, 1989). The 1991 ICMA survey shows female representation levels to be slightly higher in smaller cities with staggered council terms

(19.1%) than those using simultaneous election formats (16.9%), mostly larger cities (MacManus & Bullock, 1993).

Some, including former NOW president Eleanor Smeal (1984), believe that a majority vote (runoff) requirement retards female representation. This hypothesis is based on a scenario where a women who received a plurality when facing multiple males in the primary loses if sexist voters rally to the surviving male in the runoff. However, studies that have looked at the impact of majority vote requirements on female council representation have not found them to be detrimental (Bullock & MacManus, 1991; Fleischmann & Stein, 1987). There are two plausible explanations for this. One is that sexism is not as strong as presumed. The second is that, increasingly, women are facing other women in runoffs (MacManus, 1992).

There is also little support for the notion that scheduling municipal elections (at times unique to local government rather than having municipal elections at the time of gubernatorial or presidential elections) has any impact on women getting elected to city council (MacManus, 1992).

Political and Electoral Factors Affecting Election

Over three fourths of women councilors had no previous office-holding experience (in contrast to 40% of the women mayors). Running for council is more likely to be seen as an end in and of itself for women than for men. Women councilors are less likely than their male counterparts to see council posts as a stepping-stone to higher office. Men more often acknowledge that they run for the post to enhance their business contacts. Women are more likely to attribute their candidacies to recruitment by political party leaders than to concern for a specific issue.

Women are more prone to credit neighborhood organizations and single-issue groups for their electoral success (Welch & Bledsoe, 1988, p. 24). Women council candidates are more active than men in civic activities (Darcy et al., 1987; Flammang, 1984; Merritt, 1977). In 1977, 43% of women city councilors belonged to a political organization, 32% to professional and business organizations, 29% to service groups, 29% to youth and school service groups, 29% to cultural organizations, and 65% to at least one women's group (Johnson & Carroll, 1978).

In terms of electoral success, women council candidates do just as well as men. Welch and Bledsoe (1988) found that "the number of men and women running without opposition and winning election by a large margin were in both cases within about one percentage point of each other" (pp. 23-24). Women council candidates do not experience more difficulty in raising money or in gaining newspaper endorsements than do men (Bullock & MacManus, 1990; MacManus, 1992).

Roles and Policy Preferences

As with mayors, research comparing the policy agendas of women and men council members generally shows a fairly high level of consensus. Beck (1991) found that "almost without exception, they identify the same concerns—taxes, development, and the quality of life" (p. 105). However, women have led efforts on behalf of day care, domestic violence, and sexual assault (Boles, 1991).

Male and female council members differ on how "they perceive and respond to citizen concerns and complaints, how they gather and use information, and how they feel about political maneuvering" (Beck, 1991, p. 105). Councilors, regardless of gender, see women as more responsive to constituents (Beck, 1991).

Women and men councilors tend to criticize each other's approach to council deliberations. "Men often express frustration that women ask too many questions, while women see themselves as well-prepared and think their male colleagues are often 'winging it' " (Beck, 1991, p. 103). Welch and Bledsoe (1988) found that women devote more time to council activities than do men. Of female councilors, 56%, compared with 40% of males, reported spending more than 20 hours a week on council activities. The researchers speculate that one reason for the differential could be that more male councilors have a full-time job. If so, this gap will narrow as more women are in the workforce full-time.

■ Women in County Elected Offices

Despite the presence of thousands of counties and their control over the taxing, spending, and borrowing of billions of dollars annually, little is known about female office holding at this level. The little research that exists has focused on women on county commissions rather than on executive or judicial posts.

Women County Commissioners

Women have made significant progress in capturing slots on county commissions, but their gains lag their success at winning municipal and school board offices. Between 1975 and 1988, the number of women serving on county governing boards rose from 456 to 1,653, or from 3% of all county commission seats to 9% (CAWP, 1992).[3] As of 1988, 61.2% of the counties had no women board members (1988 ICMA survey).

Who Wins and Where?

Similar to women mayors and city councilors, females have made greater inroads in large, urban areas. A 1993 survey of the nation's largest counties (1990 populations over 423,000) found that the percentage of women on these boards was 27.5%. The propensity of women to win county commission seats in more urban, cosmopolitan areas can be observed by comparing the electoral successes of women in Florida, Georgia, and South Carolina—three states whose counties have been studied extensively. In Florida, a more urban state, the number of women commissioners jumped from 2 in 1950 to 18 in 1978 to 72 (19.5%) in 1991 (Carver, 1979; MacManus, 1991). By 1991, 56.7% of Florida's 67 counties had at least 1 female member; 25% had 2 or more, and females held a majority on six boards. In general, women fared better in counties with larger, older female populations and in counties with higher percentages of Republican registered voters (MacManus, 1991). Women also did best in counties with mixed and at-large election systems (consistent with the patterns for women city councilors).

Georgia women made far less progress than did women in Florida. Bullock (1990) has found that women rarely serve on Georgia's 159 county commissions. Between 1981 and 1989, the percentage increased only marginally from 5.1% to 5.8%, with women more successful in urban areas and in counties with mixed election systems and larger commissions. In South Carolina, the percentage of women county commissioners almost doubled between 1981 and 1989 (from 5.5% to 10.3%), still considerably below that of Florida women. Again, women did better in urban areas.

A 1988 ICMA survey found that women do best in large urban counties with more affluent, better educated populations, and appointed county administrator (manager) forms of government. DeSantis and Renner (1992) observed that "types of election systems do not have a significant impact on the number of women elected when controlling for other variables" (p. 150).

Backgrounds of Women County Commissioners

The most extensive study of the backgrounds of women commissioners found them most likely to be in the 40- to 59-year-old category (Johnson & Carroll, 1978). Three fourths had at least some college; half were in professional or technical occupations (compared with 36% of the women mayors and 37% of the council members), 65% belonged to political organizations, and 53% had held a political party position. Further, 59% of female county commissioners were Democrats, 30% were Republicans, and 11% were Independents—figures very close to their male counterparts. Women commissioners were nearly evenly split in their self-described ideology: 32% liberal, 33% middle-of-the-road, and 35% conservative. Many (63%) female commissioners were in their first elective office.

Roles and Policy Preferences

Little research beyond the Johnson-Carroll (1978) study has examined the roles and policy preferences of women county commissioners. Women commissioners (and mayors and councilors) in this study endorsed ratification of the Equal Rights Amendment, opposed a constitutional ban on abortion, and favored extension of Social Security to homemakers. Nearly half said chauvinism, stereotyping, or not being taken seriously was the biggest obstacle they experienced in office. But 40% claimed that they received higher visibility, better publicity, and special courtesies as a consequence of their gender.

More than 80% of women county commissioners perceived that they generally devote more time to the job and do better on "human relations" aspects than do men. A much higher proportion of female county commissioners than women mayors or city councilors (40% vs. 20% and 18%, respectively) had high political ambitions. Their ambitions were even higher than those of their male counterparts.

■ Women in Other County Offices

A variety of officers such as sheriff, probate judge, clerk of the court, and some type of revenue officer are elected in counties. Bullock (1990) shows a wide variation in the share of county positions filled by women in Georgia and South Carolina. In 1989, approximately three of every five clerks of court in these states were female, as were approximately half of the revenue-related elected positions in the two states and about 40% of the probate judgeships. On the other hand, neither state had a female sheriff in 1989 and women coroners, solicitors, and surveyors were extremely rare.

Bullock (1990) observes that the positions most frequently held by women are ones that have a record-keeping or bookkeeping component. He goes on to note that the positions having the greatest discretion (sheriff and solicitor) have largely been beyond the reach of women. The administrative positions to which women are most frequently elected are ones that bear some similarity to traditional female occupations. The distribution of women across offices also conforms to the expectations derived from desirability theory. Positions having greater discretion would, ceterus paribus, be more desirable than those in which the incumbent's actions are circumscribed. Sheriffs and solicitors have latitude in deciding the severity of the charges to be brought against suspected wrongdoers or whether to seek to exact a penalty.

■ Women on School Boards

Women have been more successful at winning seats on school boards than on any other elective legislative body. Data reported by the National School Boards Association (1993) show that between 1978 and 1992, the percentage of women school board members jumped from 25.7% to 39.9%.

Two long-standing notions suggest why women occupy higher proportions of school board seats. The first relates to desirability theory and posits that because school boards have traditionally been viewed as "lower in the prestige order" than other elective positions, women are more likely to run for these posts. As noted by Rada (1988),

> Compared to many other political offices, the benefits of school boards membership are slight. . . . There is no direct income (salary or wage) and little material advantage to be gained from school board membership. The degree of power is also low when compared to many other political offices. (p. 229)

More recent research (MacManus & Suarez, 1992) suggests that increased attention to all facets of education by the media and the public at-large and the higher stakes attached to controlling educational policy and expenditures have suddenly weakened the "lower prestige" or "occupational segregation" explanation for increases in female school board representation. (p. 2)

The second notion is that women are more successful at winning school board seats because their service on school boards is not threatening to men in that it is consistent with the traditional roles of motherhood and child rearing (Bers, 1978).

Who Wins and Where?

Descriptive data on school board members are collected annually by the *American School Board Journal* and the Virginia Polytechnic Institute. These surveys find women most frequently serve on school boards in the nation's geographic extremes. In 1992, 42% of the board members in the 9 Pacific states were female, as were 43.9% of the members serving in the 12 Northeastern states and the District of Columbia. Women least often served in the central states (Midwest) and western states (mountain and plains) where between 36% and 37%, respectively, of all members were female. Twelve southern states had 40.2% women on their school boards, and this region experienced the greatest increase in female board members during the decade, rising from 22.3% female in 1982. MacManus's (1991) survey of Florida school boards found that 38.1% of their members were women. In contrast, Bullock's 1989 survey of Georgia school boards found only 22.9% female, although that constituted a 10% increase for the 1980s (Bullock, 1990).

Studies of female school board representation in two southern states (Florida and Georgia) support the hypothesis that women do better in larger, more urban, more cosmopolitan settings, regardless of the local office being sought. In Florida, women do better in counties with more constituents 65 years of age or older, more Republican registrants, fewer racial minorities, and where school board elections are conducted in an at-large or mixed format (MacManus & Suarez, 1992). Although also finding that women did better when running at-large, the Georgia study differed from the Florida one in that it found a *positive* relationship with the proportion of Blacks in the electorate (Bullock, 1990).

The partisan leanings of Florida female school board members closely resemble those for women mayors, city council members, and county commissioners. There were relatively small percentages of Black and Hispanic female board members: 6.8% Black and 1.5% Hispanic.

Roles and Policy Preferences

The National School Boards Association surveys have occasionally probed male-female differences in role perceptions and policy preferences. The most recent look (Luckett, Underwood, & Fortune, 1987) found that almost three fourths of the women on school boards saw themselves as best qualified in the area of curriculum and instruction, whereas two thirds of men saw finance as their primary expertise. When asked to identify board areas in which they were most active, about two thirds of men and women identified policy making. After policy making, women said they were most active in planning and goal setting, community relations, and curriculum and instruction. Men, on the other hand, indicated they were most active in budget development, collective bargaining, facilities, and business management.

Little is known about women serving as elected school superintendents. Bullock's (1990) study of Georgia school systems found that 10% of elected school superintendents were women, up from 6% at the beginning of the decade.

■ Women of Color: Putting Things in Perspective

Since 1975, the National Roster of Black Elected Officials, which is prepared annually by the JCPES, has reported the number of African American women in public office. In that first year, the total was less than 400, with more than half serving on school boards. About one third of the total served on city councils, but only 25 were county commissioners, and just nine were mayors. By 1992, the JCPES census showed a total of 1,796 Black women in elective office. Over the 17 years, the proportion of African American women holding local office who served on school boards was down to 34% of the total, whereas the share of all Black local female officeholders who were serving on city councils had risen above 50%. The number of Black female mayors was up to 75 and Black female county commissioners to 80.

Among Hispanic females (Latinas) holding elective office in 1993, almost two thirds sat on school boards (964), 319 held municipal offices, and 127 were county officials. These are remarkable gains from the 492 Latina officeholders tallied in 1986, of whom 75 served in county government, 138 in city government, and 223 on school boards (NALEO, 1993).

By 1992, Black females held 18% of all local posts held by African Americans. In 1993, Latinas held 30.1% of all elective posts held by Hispanics. Virtually no studies have contrasted the backgrounds (socio-economic, political), role perceptions, and policy preferences of minority male and female elected officials. This leaves a big gap in our ability to discern the interactive effects of gender and race.

■ A New Research Agenda: Filling in the Missing Blanks

This overview of women in local elective office has shown that greater percentages of women holding legislative offices serve on school boards, followed by city councils, then county commissions. In terms of elected executive posts, women are most successful at winning mayoral slots and some types of elected county posts (e.g., clerk of the court, probate judge, tax collector). Fewer women have been elected school superintendent and to other county executive and judicial posts (e.g., surveyor, coroner, sheriff).

Regardless of the type of local elective office, women fare better in large, urban, cosmopolitan communities. Many are holding their first elective post. Most are better educated than their constituents and have been quite active in community organizations of all types—political, professional, and service. Many hope to run for higher office. Those in legislative positions more often win when there are more seats up for grabs and under at-large or mixed elections formats rather than single-member district formats. Other election system attributes, such as runoff requirements, term structure (staggered vs. simultaneous), or election timing, have little impact on the election of women to local office. There is also little evidence that women are disadvantaged vis-à-vis men in fund-raising or endorsements by local political parties and local newspapers.

Male-female differences in role perceptions, governing styles, and policy priorities among local elected officials *appear* to parallel many of those observed among state legislators. An extensive study of state legislators (CAWP, 1991) has found that women public officials are

- more likely to give priority to women's rights and to public policies related to women's traditional roles as caregivers in the family and society

- more active on women's rights legislation, whether or not it is their top priority
- more feminist and more liberal than men in their attitudes toward major public policy issues
- likely to have close ties to women's organizations
- more likely than men to bring citizens into the governing process
- more responsive to groups previously denied full access to the policy-making process

It is in the areas of male-female roles, governing styles, and policy priorities that *much* more research is needed. Studies contrasting the outlooks, approaches, and policy impacts of women and men in local elective office have, at best, been sporadic and based on limited databases. So, too, have been studies of minority male-female differences. As a consequence, we know very little about the interactive effects of gender and race among those seeking office at the local level.

Finally, there needs to be more research on the career paths and political ambitions of local office seekers to see if male-female ambition differentials observed some time ago persist. These kinds of studies will be much more useful if scholars are more inclusive in the types of local offices analyzed. There are, as we have shown, many local elective posts, especially at the county and school board levels, that have been underexamined.

NOTES

1. Much of the discussion in this chapter of mayors and council members comes from our article "Women and Racial/Ethnic Minorities in Mayoral and Council Positions," which appeared in *The Municipal Year Book 1993* (MacManus & Bullock, 1993, pp. 70-84). The article summarized a vast amount of literature on women mayors and council members.

2. The study did not report a separate figure for Hispanic female mayors and another for Hispanic female council members. However, a list of Hispanic mayors in 1993 supplied to us by the National Conference of Mayors shows that less than 5 of the 64 were female, meaning that most of the figure reported by NALEO are women council members.

3. To put things in perspective, the 1,653 female county commissioners in 1988 was only slightly more the 1,375 women who served in all state legislatures. Five states (Wisconsin, Illinois, New York, Michigan, and Tennessee) each had more than 100 female county commissioners, whereas 11 states had fewer than 10. The states with the most counties (Texas with 254; Georgia with 159) both had relatively few female commissioners—30 in Texas and 28 in Georgia (Bullock, 1990).

REFERENCES

Beck, S. A. (1991). Rethinking municipal governance: Gender distinctions on local councils. In D. L. Dodson (Ed.), *Gender and policymaking: Studies of women in office* (pp. 103-111). New Brunswick, NJ: Center for the American Woman and Politics.

Bers, T. H. (1978). Local political elites: Men and women on boards of education. *Western Political Quarterly, 31*, 381-391.

Bledsoe, T., & Herring, M. (1990). Victims of circumstances: Women in pursuit of political office. *American Political Science Review, 84*, 213-223.

Boles, J. K. (1991). Advancing the women's agenda within local legislatures. In D. L. Dodson (Ed.), *Gender and policymaking: Studies of women in office* (pp. 39-48). New Brunswick, NJ: Center for the American Woman and Politics.

Bullock, C. S. III. (1990, November). *Women candidates and success at the county level.* Paper presented at the annual meeting of the Southern Political Science Association, Atlanta, GA.

Bullock, C. S. III, & MacManus, S. A. (1990). Voting patterns in a tri-ethnic community: Conflict or cohesion? The case of Austin, Texas, 1975-1985. *National Civic Review, 79*, 23-36.

Bullock, C. S. III, & MacManus, S. A. (1991). Municipal electoral structure and the election of councilwomen. *Journal of Politics, 53*, 75-89.

Carroll, S. J. (1985). *Women as candidates in American politics.* Bloomington: Indiana University Press.

Carver, J. (1979). Women in Florida. *Journal of Politics, 41*, 941-955.

Cassel, C. A. (1985). Social background characteristics of nonpartisan city council members: A research note. *Western Political Quarterly, 38*, 495-501.

Center for the American Woman and Politics. (1991). *The impact of women in public office: Findings at a glance.* New Brunswick, NJ: Author.

Center for the American Woman and Politics. (1992). *Fact sheet: Women in elective office 1992.* New Brunswick, NJ: Author.

Darcy, R., Welch, S., & Clark, J. (1987). *Women, elections, and representation.* New York: Longman.

DeSantis, V., & Renner, T. (1992). Minority and gender representation in American county legislatures: The effect of election systems. In W. Rule & J. F. Zimmerman (Eds.), *United States electoral systems: Their impact on women and minorities* (pp. 143-152). New York: Greenwood.

Flammang, J. A. (1984). Filling the party vacuum: Women at the grass-roots level in local politics. In J. A. Flammang (Ed.), *Political women: Current roles in state and local government* (pp. 87-114). Beverly Hills, CA: Sage.

Fleischmann, A., & Stein, L. (1987). Minority and female success in municipal runoff elections. *Social Science Quarterly, 68*, 378-385.

Hill, D. B. (1981). Political culture and female representation. *Journal of Politics, 43*, 151-168.

International City Management Association. (1981). *Form of government survey* [Data set]. Washington, DC: Author.

International City Management Association. (1988). *Form of government survey* [Data set]. Washington, DC: Author.

International City/County Management Association. (1991). *Form of government survey* [Data set]. Washington, DC: Author.

International City/County Management Association. (1993). *The municipal yearbook.* Washington, DC: Author.

Jennings, J. (1991). Black women mayors: Reflections on race and gender. In D. L. Dodson (Ed.), *Gender and policymaking* (pp. 73-79). New Brunswick, NJ: Center for the American Woman and Politics.

Johnson, M., & Carroll, S. (1978). Statistical report: Profile of women holding offices, 1977. In Center for the American Woman and Politics (Ed.), *Women in public office* (2nd ed., pp. 1A-65A). Metuchen, NJ: Scarecrow.

Joint Center for Political and Economic Studies. (1992). *National roster of Black elected officials 1992.* Washington, DC: Author.

Karnig, A., & Walter, B. O. (1976). Election of women to city councils. *Social Science Quarterly, 56,* 605-613.

Karnig, A. K., & Welch, S. (1979). Sex and ethnic differences in municipal representation. *Social Science Quarterly, 60,* 465-481.

Luckett, R., Underwood, K. E., & Fortune, J. C. (1987, January). Men and women make discernibly different contributions to their boards. *American School Board Journal,* pp. 21-41.

MacManus, S. A. (1981). A city's first female officeholder: "Coattails" for future office seekers? *Western Political Quarterly, 34,* 88-99.

MacManus, S. A. (1991). Representation at the local level in Florida: County commissions, school boards, and city councils. In S. A. MacManus (Ed.), *Reapportionment and representation in Florida: A Historical Collection* (pp. 493-538). Tampa: University of South Florida, Intrabay Innovation Institute.

MacManus, S. A. (1992). How to get more women in office: The perspectives of local elected officials (mayors and city councilors). *Urban Affairs Quarterly, 28,* 159-170.

MacManus, S. A., & Bullock, C. S. III. (1989). Women on southern city councils: A decade of change. *Journal of Political Science, 17,* 32-49.

MacManus, S. A., & Bullock, C. S. III. (1992). Electing women to city council: A focus on small cities in Florida. In W. Rule & J. F. Zimmerman (Eds.), *United States electoral systems: Their impact on women and minorities* (pp. 167-181). New York: Greenwood.

MacManus, S. A., & Bullock, C. S. III. (1993). Women and racial/ethnic minorities in mayoral and council positions. In *The municipal yearbook 1993* (pp. 70-84). Washington, DC: International City/County Management Association.

MacManus, S. A., & Suarez, R. (1992). Female representation on school boards. *Political Chronicle, 4,* 1-8.

Merritt, S. (1977). Winners and losers: Sex differences in municipal elections. *American Journal of Political Science, 21,* 731-744.

National Association of Latino Elected and Appointed Officials Educational Fund. (1993). *Hispanic elected officials top 5,000 mark nationwide: Almost one out of three officials is Latina in 1993.* Washington, DC: Author.

National School Boards Association. (1993). *School boards: The past ten years.* Alexandria, VA: Author.

National Women's Political Caucus. (1993). *Fact sheet on women's political progress.* Washington, DC: Author.

Rada, R. D. (1988). A public choice theory of school board member behavior. *Educational Evaluation and Policy Analysis, 10*, 225-236.

Rinehart, S. T. (1991). Do women leaders make a difference. In D. L. Dodson (Ed.), *Gender and policymaking: Studies of women in office* (pp. 93-102). New Brunswick, NJ: Center for the American Woman and Politics.

Robinson, T., & England, R. E. (1981). Black representation on central city school boards revisited. *Social Science Quarterly, 62*, 495-502.

Smeal, E. (1984, June 28). Eleanor Smeal report. *Eleanor Smeal Newsletter*, p. 1.

Welch, S., & Bledsoe, T. (1988). *Urban reform and its consequences*. Chicago: University of Chicago Press.

Welch, S., & Herrick, R. (1992). The impact of at-large elections on the representation of minority women. In W. Rule & J. F. Zimmerman (Eds.), *United States electoral systems: Their impact on women and minorities* (pp. 153-166). New York: Greenwood.

Welch, S., & Karnig, A. K. (1979). Correlates of female office-holding in city politics. *Journal of Politics, 41*, 478-491.

Part III

Environments

10 Developing Rape Programs and Policies Based on Women's Victimization Experiences: A University/Community Model

LYN KATHLENE

Rape and the college woman suddenly became a major media event by the beginning of the 1990s. In 1985, *Ms.* magazine released its extensive survey results of 6,159 students on 32 college campuses in the United States and found that one in four college women have been victims of rape or attempted rape (Warshaw, 1988). Moreover, these attacks did not fit the stereotype of a stranger lurking outside the library, waiting for a student walking back to her dorm late at night. No, 85% of these rapes were committed by a male acquaintance (Koss, Dinero, Seibel, & Cox, 1988). The message was that young women need not fear the unknown as much as the known.

Yet in spite of society's growing acknowledgment of these gender crimes, a woman *assaulted by a man she knows* continues to face tremendous emotional, social, and legal barriers to resolving her victimization. Women are doubly victimized—by their attackers and, subsequently, by society. Although feminist critiques of patriarchy provide a metalevel theory of women's subordinate position in society and help us understand how individuals encode and respond to rape victims, combining feminist theory with urban theories (of social learning, territory, and service delivery) provides insights capable of transforming institutions.

AUTHOR'S NOTE: Five undergraduate students were instrumental in collecting the data for this research. Many thanks to Karin Fisher, Lori Long, Anne Morgan, Karen Rexing, and Michele Vakili for their unselfish dedication to making Purdue University a better place for women.

This chapter provides an example of how we can begin to systematically make our communities safer for women. Following are the results of a two-semester student research project, directed by the author, that provided the university and community with a comprehensive view of a broken system, articulated the very real pain of our women students who had been assaulted, and made policy recommendations for overcoming institutional barriers that artificially separated the university campus from its community.[1] Although the impetus for the study began with students at a large midwestern public university concerned about rape on campus, it soon became apparent that defining the problem of women's safety in terms of artificial institutional boundaries ("on-campus") was one of the problems. The severe incongruence between the nature of the crime (it could happen anywhere) and the institutional response (first and foremost defined by their jurisdictional authority) results in an ineffectual, if not harmful, service delivery system.

■ Theoretical Background

A Feminist Critique

Private and public colleges across the nation have been forced to examine sexual assaults on their campuses (Gibbs, 1991). Partly in response to the *Ms.* study, "Take Back the Night," marches all across American campuses began to symbolize more than a feminist-fringe cry against women's victimization by strangers lurking in the shadows. Women attending Brown University, fed up with their university's response to rape victims, began writing and carving the names of rapists on the walls of bathrooms across campus. Young women were learning that female sexual victimization was both widespread and widely ignored by campus and community authorities.

During this time, experimental psychologists were collecting and analyzing data from their undergraduate psychology students regarding attitudes toward hypothetical rape victims based on adherences toward traditional sex roles and/or beliefs in rape myths. Study after study found that men and women who adhered to traditional notions of sex roles gave more credence to rape myths (see, for example, Jenkins & Dambrot, 1987; Krahe, 1988). Also, these attitudes were correlated with victim blaming. Even more disturbing was the finding that 30% of college men claimed that they would rape if they thought they could get away with it and 1 in 12 college men admitted committing acts that meet the legal definitions of rape (Warshaw, 1988).

Although these psychological studies, themselves, are alarming, understood through the insights of feminist theory, the link between individual constructions of the social and the social construction of individuals takes on terrifying proportions. That sexual violence against women cannot (must not) be explained away as deviant behavior by *some abnormal* males is a core theme developed in critiques of patriarchal societies (see, for example, Brownmiller, 1975). Despite recent "gains" in educational and work opportunities, the continued social construction of the feminine as the "other"—as "inferior"—pervades our communities at all levels, expressed through physical or emotional battering by a domestic partner (Pagelow, 1984); occupational sex segregation or the infamous "glass ceiling" prevalent throughout entrepreneurial and corporate America (Kemp, 1994); or the daily public objectification of the feminine body to sell cars, beer, and entertainment. Taken together, this culture serves to subordinate and oppress women in every aspect of social life.

Why a woman does not report her attack, especially when sexually assaulted by a man she knows, is largely conditioned by these singular patriarchal lenses she has been given to understand her victimization. But when she *does* seek help from friends, family, police, administrators, or counselors, they too may view her assault through similarly distorted lenses, adding to her emotional trauma and confusion.

An Urban Perspective on Service Delivery

This social construction of the rape victim is confounded by the institutional design of service delivery, especially in urban areas divided into autonomous legal entities. Coordination occurs only to the extent that agencies refer clients to the "appropriate" agency. Thus a student raped at a fraternity house logically would contact the campus police, but she will learn that the crime must be reported to the city police because the house is not university property.

Moreover, student pressure to create or improve services is realistically limited to campus providers only, even though it often may be off-campus agencies that are responsible for providing services to students. So although students may effectively pressure the campus police to be more responsive—for example, to have only female officers interview a rape victim—such lobbying and influence is absent with city and county law enforcement. Despite the fact that most students will be raped off-campus (due to limited housing on university property and the situation surrounding most acquaintance attacks), administrative responses to students' social problems are curiously circumscribed by boundaries based on the property lines of their campus rather than on the social life of their students.

The university, city, and county responses—although all operating within the larger confines of a hierarchical and patriarchal society that serve to devalue women (Ferguson, 1984)—are also driven by institutional pressures to protect and therefore enhance their own authority (Herson & Bolland, 1990). The city police find they must service students but may not define this group of people and their problems as a central duty. City police chiefs probably have little incentive to revamp their procedures to respond more effectively to a population that "belongs" to another law enforcement agency (the campus police). Even when agencies finally respond to problems, the solutions are likely to be less than optimal because most institutions change only incrementally. Old solutions will be favored over new and creative solutions (Brewer & deLeon, 1983), thereby enhancing the status quo and potentially compounding the existing problem.

Finally, the implementation of policies responsive to sexual assaults, however well designed at the administrative level, will likely fail without careful feminist training of the people a rape victim may contact for help. Whether this be her friends (likely to be young, inexperienced women similar to herself), a dorm counselor, the clergy, or the police, the socially scripted roles of male and female sexual behavior that bombard us on a daily basis will undermine change unless they are explicitly addressed in the implementation of sexual assault policies.

As long as women's experience is effectively hidden (read: *suppressed*) by these social, legal, and institutional constructions of women's sexual victimization, women will continue to be doubly victimized—by the physical assault and by the atmosphere that subsequently silences them. To reconstruct women's experience and deconstruct our misogynist environment, we must begin to hear how women are being silenced.

■ Hearing the Silenced Voices

Nearly all the survey respondents (94%) were victims from 1988 to 1991. Because the typical undergraduate student graduates in 4 years,

TABLE 10.1   Women Raped or Attacked While Attending the University, 1991 Survey (*N* = 50)

Type of assault	
Rape[a]	62%
Attempted rape	22%
Other[b]	16%
Location of assault	
On-campus	35%
Off-campus, within city limits	55%
Outside county	10%
Relationship to assailant	
Stranger	29%
Acquaintance	69%
Of these (*n* = 34)	
Knew one month or less	36%
Average length of time	4.9 months
Mean age of woman when attacked	19.1 years
Class status of woman when attacked	
Freshman, 1st semester	33%
Freshman, 2nd semester	27%
Sophomore	10%
Junior	12%
Senior	4%
Graduate	4%
Other[c]	10%
Mean age of male assailant(s)	21.4 years
Student status of assailant(s)	
Not a student (at this university)	19%
Student at the university	81%
Of these (*n* = 35)	
Fraternity member	54%
Number of male assailants	
One	84%
Two or more[d]	16%
Of these (*n* = 8 attacks)	
Victim knew assailants	75%
Fraternity members	50%

a. Includes vaginal rape and forced fellatio.
b. Includes bondage and beating, molestation, and physical assault with objects.
c. Includes staff and alumni.
d. Women attacked by multiple assailants did not form their own cluster type. These women had a variety of responses, showing up in five of the seven cluster types, which means that the number of assailants was not the defining characteristic of the rape or their postattack help-seeking behavior.

this predominance of fairly recent attacks was not surprising. Table 10.1 reports the descriptive statistics of the sample.

A typology of victims' responses was developed through cluster analysis. We were interested in women's responses to the situational characteristics of the attack, whom they told about the attack, and what types of reactions they received from their social and community network.[2] The seven types represent profiles of women who shared similar experiences in terms of the circumstances of their attacks and their help-seeking behavior afterward. The characteristics common to the women in each type are presented first. The profiles are discussed in descending order, from the best- to the worst-case scenario, based on both the victim's help-seeking behavior and the responses she received from the people she contacted. This approach provides an orderly view of which circumstances surrounding a rape are socially and institutionally construed as legitimate and which are not. In this way, we are able to define when and how the system, including friends and family as well as service providers, fails women who have been sexually assaulted.

Type 1: Reported—Attempted Rape by a Stranger ($n = 7$)

- place and time: attempted rape outside, at night
- assailant: one stranger, believed to be about 22 years of age
- victim: 19 years old, second-semester freshman
- postrape experience: averaged 3.4 contacts
    1. friend, within hour of attack
    2. police, within hours of attack
    3. dorm adviser, within 24 hours of attack
    4. family, within 24 hours of attack

The best postattack responses were experienced by women who successfully escaped an unknown rapist—the man attacked a woman unexpectedly while she was alone outside at night either on- or off-campus. The on-campus attacks happened in dormitory parking lots. The off-campus attacks typically occurred close to her apartment. She immediately contacted a sympathetic and supportive female friend who encouraged her to call the police. However, if she told her boyfriend, he reacted angrily wanting to "get the guy" *or* he "had trouble understanding" her victimization because she had successfully escaped.

Half of the women attacked off-campus called the campus police first and had to be referred to the city police. The police responded well (relative to the other types). They encouraged a woman to file a formal report and to press charges; however, the assailant was never found. Although the police took her claim seriously during the initial contact, half of the women felt that the police were either "too businesslike," "dropped the case too quickly," or did not keep them "informed sufficiently."

It is no coincidence that women who experienced the stereotypical sexual assault—by a stranger, late at night, outdoors—received the most supportive responses from both their social and community networks. Equally important is that women attacked (unsuccessfully) by rapists who are strangers readily recognize themselves as victims, too.

Type 2: Reported—Rape by a Very New Acquaintance ($n = 6$)

- place and time: fraternity or assailant's apartment, at night
- assailant: acquaintance, known for less than 1 month, 20 years old
- victim: 18 years old, a first-semester freshman
- postrape experience: averaged 5.3 contacts
    1. friend, within 24 hours after the attack
    2. police, 6 days after the attack
    3. hospital, 14 days after the attack
    4. family, 2 months after the attack
    5. dean of students, 6 months after the attack
    6. psychologist, 10 months after the attack

This group of women made more contacts than any other type. Most of the women were advised by the police (but sometimes by a counselor or teacher) not only to report the attack but also to press charges. There was a tendency for the police to discourage some women from filing a report if they were not willing to press charges or if the case appeared "weak." One woman who brought charges against the man who raped her at a fraternity party explained the circumstances:

> I was at a party with some friends. We were talking to a few guys in one of the rooms. We were trying to dodge the crowds. My friends and the rest of the guys left the room to get refills of their drinks. The guy went to the door, locked it, and attacked me.

Although she insisted on filing a report with the campus police, she was not encouraged to press charges "because of the circumstances . . . it wouldn't be a very good case." Nevertheless, she persisted, and he was charged with rape and battery. Ultimately, the case was dropped because, as she reports, "I had agreed to go to the party with my friends, so it was not all his fault." This assessment appears to be her own self-blaming, either encouraged, or at least not discouraged, by the treatment of her case in the legal process. She was told that hers is not "a very good case" because she had willingly gone to the party, was drinking alcohol, and trusted a man she barely knew. Simply put, she now knows that she should have known better. The outcome of this stereotypic social and legal process? The victim learns a lesson; the rapist is free to "teach" again.

Type 3: Unreported—Rape by a Very New Acquaintance ($n = 4$)

- place and time: assailant's room, at night
- assailant: known for 1 month, 19 years old, fraternity member
- victim: 18 years old, first-semester freshman
- postrape experience: averaged 3.75 contacts
  1. friend, 2 days after attack
  2. hospital, 5 days after attack
  3. family, 3 weeks after attack
  4. dorm adviser, 3 months after attack
  5. psychologist, 3 months after attack

Raped within the first 8 weeks of becoming a university student, these young women knew their attackers as recent "friends." Half of the women were attacked at a fraternity party; half were attacked in the man's dorm room. Although there was no indication that the women or men were drunk, several mentioned that they had one or two drinks. One woman wrote,

> My friends and I had met "Ben" a couple of days earlier. We all ended up at a party together on a Friday night. I had one wine cooler to drink. He had a couple of beers at most—he was not drunk. He invited my friends and me back to his room to watch a movie. We went, myself and two female friends. My two friends started to get tired and wanted to leave. I hadn't seen the movie and wanted to see how it ended. Ben said he would be happy to walk me home after the movie. So my friends left. Not 10 minutes later he was all over me. I told him NO, kicked and screamed . . . Ben was a big guy, about six-one and over 200 pounds. I'm five-five and 125 pounds.

Female friends had a mixed response. Some were shocked, supportive, and sympathetic, but none encouraged reporting the rape to the police. Others blamed the woman for her poor judgment, and one friend convinced the rape victim that she had not been raped. Two of the women in this group got medical attention but not until an average of 5 days had passed. The other two women did not go to the hospital because they were "too humiliated," were "talked out of it" by a friend, or felt that "too much time had elapsed" at the point that they considered having a medical exam. Although family members responded with concern and support, a woman did not share her victimization with them until an average of 3 months later. One woman, who came from a religious home, was afraid to tell her family that the first time she drank a beer she also got raped (and lost her virginity). The emotional trauma is evident among this group of women: Most entered psychological counseling within 3 months. One woman attempted suicide 2 weeks after the attack. Another woman summarized her reaction to the rape that had happened 1.5 years ago:

> I was a freshman. It was the first time I drank and I thought it was my fault he did it. . . . I didn't want to talk about it. I really didn't think it would affect me in the long run, but as of today, I don't think a day has gone by that I haven't thought about it.

None of the women went to the police. Reading their statements, it is clear why these women did not report their rapes. All were away from home for the first time. Within the first 3 months living as an independent adult, the woman found herself betrayed, attacked, and raped by a new friend. She questioned her own judgment, and her equally young and inexperienced women friends questioned her judgment. She was confused, humiliated, and lacked a social network that could help her understand her victimization.

Type 4: Unreported—Rape and Continuing Threats by a New Acquaintance (*n* = 5)

- place and time: assailant's room or apartment, at night
- assailant: acquaintance, known for 2 months, 21 years old; half were from another university
- victim: 18 years old, first-semester freshman
- postrape experience: averaged 1.6 contacts
  1. friend, within 24 hours of the attack

These women told very few people about the assault. Most of the women received very negative or accusatory responses from the friend or friends that they told. One woman was warned by a female friend not to report the rape because she would "face a lot of trouble with the fraternity and Intra Fraternity Council." Other friends responded in disbelief that the attack really occurred or blamed the victim for "leading him on." One woman was raped by a prominent male athlete and was too scared of him and the publicity that would follow if she reported the rape. Most of the victims were harassed and threatened afterward by their assailant, too. Typical of the fear and rejection these women experienced was expressed by one women who found no support from either her friends or the clergy:

> I really wish I could find someone who cared, especially on the nights he calls in threats to me or shows up at my dorm to threaten me. I get scared and know it was all my fault anyhow. I have to learn to beat him up or something.

The tendency of these women to blame themselves is understandable, given the paucity of appropriate responses from their social circle. Moreover, living with postrape harassment and threats, these young women continued to be emotionally battered by their assailants. It is possible that the rapist verbally demeans her, redefines the attack, and threatens her as a way to control the event. She, without a social support group, fears him and blames herself—a psychological response common among physically and emotionally battered women.

It appears that the first contact (i.e., the friend) was particularly crucial for these women. It is unclear what dynamic produced unsupportive friends *and* postrape threats, unless the assailant finds out that the victim does not have a social support network. This is certainly possible, because so-called friends who ridicule a rape victim's story are unlikely to feel bound to confidentiality. Perhaps this lack of social validation of her attack even helps spur the rapist's continued contact and harassment. She is an easy target; no one believes her. For these men, the continued threat of violence can act as an extension of the rape.

Type 5: Unreported—Rape by an Unknown Man ($n = 10$)

- place and time: assailant's fraternity/dorm room or outside, at night
- assailant: unknown student, believed to be 22 years old

- victim: 18 years old, a second-semester freshman
- postrape experience: averaged 1.3 contacts
    1. friend, 2 days after the attack

Unlike the women who successfully fended off the stranger that attacked them, women who experienced a completed rape by a stranger told very few people. The label "unknown student" was chosen to convey that the attacker was not qualitatively the same as the "scary" stranger lurking in the bushes, but rather was "just another" male student attending the same party as the woman he chose to attack. Women were raped either in the upstairs room of a fraternity house or an off-campus residential house, or outdoors in close proximity to where the party was being held. Many, but not all, of the women mentioned that they had been drinking and feared that reporting the rape would result in public humiliation for them and, most important, for their family. Additionally, all of these women stated in one way or another that they were too embarrassed to talk to a stranger (a police officer) about the rape.

Most friends did not react negatively to the news, but few were able to provide the needed emotional support or knowledge about what to do next. One woman who was raped by a "fraternity brother" at a house party said:

> I told my friend because I didn't know if I should tell my boyfriend. She didn't have a very strong reaction because the same exact thing happened to her—drunk at a party and raped while passed out. My boyfriend threw a fit and was angry at *me* until he realized I was the *victim* [emphasis in the response].

And, once again, women who told their boyfriends found a dearth of appropriate responses. Boyfriends were uniformly angry, either at the woman or the rapist, and were unable to provide the woman with any help. The one exception, a lone male friend (platonic), contacted the police to report the rape of his friend but found that the police would not even "log in the call" without her filing a formal report, which she was too afraid to do. Yet her rape was just one of several that her friend knew had happened at this particular fraternity.

Indeed, the most disturbing connection between the women in this type was the frequency with which they found out that their friends had had similar experiences and the belief that nothing could be done about it. One woman bluntly stated:

Here at [this university] it seems that no one gets punished for it. I know many girls who reported it but just couldn't prove it. I guess that was a big part of why I didn't. . . . I guess if I thought they'd go to jail or something would be done, maybe my decision would've been different.

Although city and campus police records would show few women have ever reported being raped, the preceding statement suggests that women attempt to report the rape but are dissuaded from doing so before it becomes part of the formal record. As we saw in Type 1, even among the women who did successfully report their attack to the police, several mentioned the lack of support and poorly framed remarks they received from the officer.

Type 6: Unreported—Rape by a Well-Known Acquaintance ($n = 8$)

- place and time: assailant's room at a fraternity or dorm, at night
- assailant: acquaintance, known for 6 months, 20 years old
- victim: 19 years old, second-semester freshman
- postrape experience: averaged 1 contact
  1. friend, 5 months after the attack

Prior to the attack, none of the women were sexually involved with these well-known male friends. Women did not talk about the assault to anyone at all or not until many months later because, as one woman put it, "He was a friend. Everyone knew and liked him. I was afraid of getting called a liar and what people and my boyfriend would say." Unlike Type 4, all these women recognized their victimization and did not blame themselves. Nevertheless, their conceptualization of how others would respond to them effectively controlled their help-seeking behavior. One woman, who sought counseling from the Dean of Students's office, could not bring herself to directly tell the counselor about the rape. In response to her roundabout attempt to reveal the attack, the counselor focused on alcohol abuse, never asking the woman if she had been sexually assaulted by this close friend. Of course, the woman never told the counselor and therefore never received the help she was seeking. In this attack/response type, we see women who knew they were not to blame, but the close friendship they had had with their attacker made it extremely difficult to talk about the rape.

When a woman did share her experience with a friend many months later, she found strong emotional support. This friend, however, was not one of the "crowd" that the rapist hung out with. In fact, given the disbelief and outright backlash that women in Type 4 experienced after telling their friends in the social group that included the attacker, women in this type appear to have had good reason for assuming they would receive very negative reactions from their friends. Although much time had passed before they sought help, they wisely chose to tell a friend who had no emotional investment in disbelieving their story.

Type 7: Unreported—Rape by a Boyfriend (*n* = 6)

- place and time: at his or her place, at night, may include physical assault
- assailant: acquaintance, known for 9 months, who was 21 years old
- victim: 21 years old, at least a junior
- postrape experience: averaged 2 contacts
  1. friend, within 24 hours of the attack
  2. police, 3 days later

Most of these women were attacked by their boyfriends. As frequently occurs in domestic violence cases, most of these women were physically assaulted and then sexually assaulted. One 18-year-old woman "punched several times before [being] punch[ed] . . . in the temple" was knocked unconscious and then raped. When she contacted the campus police to report the assault, she was told it was "too much trouble to file a report without pressing charges," which she was unwilling to do because she was "so confused and . . . thought [she] still loved him"—a common reaction of women in abusive relationships. Most disturbing in this woman's account is the reaction she "received from every male authority figure—police, dorm director—[who] treated me as if it were my fault. I didn't want to continue getting drilled for information while my assailant was left to himself."

Others in this type demonstrate that even well-educated women accept the cultural myth that previous consensual sex gives the partner sexual property rights to her body. After being raped by her boyfriend and calling it a rape when telling her friends and her ex-boyfriend about it, this 25-year-old graduate student ultimately understood her type of victimization as one that cannot be overcome:

> The thing that I think is most helpful to know here is that radical feminist ideology, rape education, and criminology training had *no effect* on the

fact that *nothing* (except being dragged kicking and screaming) would have persuaded me to file any kind of report. I knew him; I was there; it was my fault (deep down, no amount of ideology can change this heart response); I had to live with it [her emphasis].

■ Understanding Help-Seeking Patterns

Although previous research has examined students, police, and rape counselors' beliefs in rape myths and the degree to which they blame the victim, and case studies have helped us understand the multiple reactions that women experience (Kelly, 1988; Roberts, 1989), the survey cluster types suggest that *both* the contacts a victim makes and the victim herself are influenced by rape stereotypes that can interact in reinforcing ways. The victim's perceptions and definitions of the attack interact with the responses she receives from her social and community network. This is particularly evident for women who experienced an attempted rape by a stranger—Type 1. These women recognized themselves as legitimate victims; immediately contacted friends, police, dorm advisers, and family; and, in return, received the most consistently positive and supportive reactions to their attack. The stereotype that women are attacked by bad men they do not know and that a good woman will fight off her attacker (attempted rapes = a successful good woman) fits the situational characteristics of this type. It is not surprising that these women and their contacts all recognized the victimization and responded appropriately.

In contrast, a college woman beaten and raped by her boyfriend, Type 7, is unable to disentangle her emotional attachment in the relationship, the cultural script that gives a man unlimited access to her body, and the physical and sexual assault. Especially important here is evidence that battering happens in dating relationships and that the appropriate responses must be grounded in the realization that these women are victims of domestic abuse, too.

In between these two extreme types are victim help-seeking patterns and responses that vary largely in accordance to the circumstances of the rape. When a rape happened at a party and/or when alcohol was involved, women, although usually recognizing their victimization, also bought into the myth that they were responsible for being raped, thus effectively limiting their ability to seek help. Equally important was the poor responses these young women received from their women friends. In the worst cases, the friend(s) did not believe the woman; in the best cases, the response was one of sympathy but helplessness to stop such attacks against women.

Perhaps some of the most useful information to arise out of the cluster types was the different help-seeking patterns that arose in women attacked by new acquaintances. Women in Type 2 immediately understood their victimization and sought out help, including pursuing formal charges against the assailant. However, with the lack of encouragement from the police, these women gradually turned the experience inward and blamed themselves for poor judgment. Again, part of this response can be understood in terms of the gendered interpretations we put on alcohol usage, but what is most disturbing is to see the power of this construct and its pervasiveness in society: Even a woman who sought help quickly and did not blame herself was eventually convinced that she played a role in her own rape.

Unfortunately, not all women attacked by a new acquaintance were able to process the assault as anything but their own fault. The very young and inexperienced women (both socially and sexually) in Type 3 could not report the rape to the police for fear of how attending a party and drinking would be viewed. They were deeply ashamed and humiliated and found little help from their equally inexperienced friends.

■ A Direction for Change

Underreporting of rapes is a nationwide problem. By listening to women who have been victimized, we learn that their low reporting rate is linked to the cultural script that blames them for their own victimization and that friends' and even service providers' responses reflect this same misogynist viewpoint. If we focus our efforts at getting women to report sexual assaults—a typical police response to complaints about misleading statistics—we will be overlooking the most important component of their postassault experience: their friends. Education programs need to go beyond warning women about dangerous situations and men about their criminal liability to including a feminist analysis of our rape culture (Fain, Richardson, & Wemmerus, 1992). For if most women seek help first and only from a friend, education efforts need to focus on the emotional and traumatic effects of rape for *both* the victim and her friends in whom she confides. In short, we need to educate women as potential victims *as well as* potential friends of victims. Because young women are at the greatest risk of being assaulted by an acquaintance (see Table 10.1; Warshaw, 1988), this education process should ideally happen early—in human development classes in high school, followed up by a mandatory freshman rape education program.

Another especially disturbing finding was the machismo reaction of a boyfriend. The woman's need to be cared for becomes overshadowed by his need to be protected from himself. Because the woman wants to protect her boyfriend, his threat to beat up or even kill the rapist can silence her. She downplays the event to diffuse his anger by denying herself the choice of reporting the rape to the police. Educational efforts aimed at men need to go beyond stressing the criminal consequences of would-be rapists, to focus on how men who do not rape contribute to the rape culture and women's victimization by responding with threats of violence.

The full range of community and campus resources must be coordinated to enhance service response. For example, it should be mandatory that police (city or campus) contact a rape counselor as soon as a rape victim arrives. Because many students who have been raped do get medical help at the campus hospital, it is also imperative that those who may be the first professional contact respond appropriately. Campus health providers need to counsel the women about their options, provide them with in-depth literature about acquaintance rape, and verify their victimization immediately. Although the woman may be accompanied by a woman friend, she will probably not have a feminist understanding of rape, which is crucial for empowering the victim, especially in a legal system that actively works against a woman attacked by someone she knows. A fully trained rape victim advocate (often available through local crisis centers) can provide help in ways that police (even police who have received victim sensitivity training) cannot, by providing information about her legal options and emotional support to understand her victimization from a feminist perspective on societal violence against women.

In any university or urban community, there will be a variety of psychological services available that rape victims may contact. These will range from trained professionals to untrained peer counselors to religious organizations. Our research found that most counselors were unaware of the variety of rape counseling services available on the campus and in the community. Particularly disturbing was the lack of training that peer counselors, such as residence hall advisers, received. Although police educated the student counselors about legal reporting procedures, most peer counselors lack knowledge about the role of alcohol in sexual assaults and the existence of rape shield laws.[3] Similarly, the clergy of different religious organizations were equally uninformed in these matters. Educational outreach must be expansive and creative to include a variety of people likely to be contacted by a victim seeking counseling. Additionally, sexual assault counseling services need to be centrally coordinated and the information widely disseminated to provide victims with a choice of appropriate services.

Traditional sex role expectations ultimately place women in the impossible position of controlling all sexual encounters—hence the need to reeducate people. But campus and city police also need to redefine their jurisdictional boundaries based on whom they serve rather than where they serve, because decentralized and uncoordinated service delivery compounds the problems of social bias against women. Rape victim advocates from community organizations not only should be granted unconditional access to campus (they currently must have a faculty/staff sponsor to conduct official business) but should be considered a regular resource person called by the police or hospital personnel when handling a sexual assault.

In tracking sexual assaults, communities and campuses must include multiple indicators, rather than solely relying on crime statistics produced by the police. Only rapes that have had a formal police report filed become a crime statistic; yet most women do not contact the police, and the few who do are usually not encouraged (either overtly or in more subtle ways) to file a report. Pronouncing that "more women should report the crime to the police"[4] has not worked in the past, nor, given women's experiences, will it work in the future. But women do contact other "authorities," from psychological counselors on- and off-campus to the clergy to resident hall advisers to local community crisis lines. In our society, often, it is numbers, not discourse, that persuades and legitimates a problem. For service providers and women in the community to "see" the problem, a systematic and inclusive data collection system needs to be implemented to enable the development of a more responsive system.

Furthermore, although police insist on a formal interview and report to help protect the accused from false charges, the police fail to acknowledge that the lack of an informal reporting procedure protects the criminal actions of an entire rape culture, such as that embodied in the fraternity structure (Martin & Hummer, 1989). The use of an additional alternative reporting system would enable the police and the university to pinpoint which fraternities have a pattern of victimizing women—an important first step in building a case against the offending house.

■ Conclusion

Ultimately, this study points out that the *reality* of women's responses to their victimization must be incorporated into educational programs, used when formulating public policy, and serve as a basis for legal reform and police procedures if we hope to create women-sensitive communities.

If individual women are judged as mainly, or even partly, to blame for their own victimization, the social and institutional conditions that support gender violence will not be addressed. To develop sensitive, comprehensive campus and community responses to gender violence, we need to document how our campuses and communities are helping and harming women. Ultimately, each campus needs to do a program evaluation, even where there is no formal program. To overcome the patriarchal construction of existing institutional responses, the evaluation must be explicitly feminist and unabashedly grounded in women's experiences (Harding, 1991; Reinharz, 1992). Only through incorporating the voices of women who have been assaulted will a new generation of policies and programs be created that serve to respect and empower women.

NOTES

1. Through interviews, analysis of documents and previous surveys of targeted groups, my class of 46 students identified what services, on- and off-campus, were available, which were being duplicated and what new services were needed. In addition, five students met with me to design and organize the administration of a four-page survey directed at women who had been attacked or raped while attending the university. The number of returned surveys ($N = 50$) does not represent the proportion of women who have been attacked. Rather, the purpose of a target survey was to learn the about the range of reactions women experienced after their attacks.

2. Forty-six cases were cluster analyzed based on 8 situational variables and 11 postattack response variables. Four cases, with extensive missing data, were removed prior to the cluster analysis. The 8 situational variables were type of attack, number of assailants, relationship between attacker and victim, length of time victim knew assailant, age of victim, class of victim, time of day the assault occurred, and race of victim and assailant. Eight response variables were dichotomous yes/no victim contacts made to the police, hospital, hotline, friend, family, dorm adviser, clergy, psychologist. Three more response variables were the average number of contacts made, whether a medical rape kit was done, and whether anyone discussed the option of filing a formal police report.

3. Rape shield laws vary in their specific content and implementation; however, the general purpose is to protect a rape victim from having her past sexual conduct "put on trial." No rape shield law completely prevents the introduction of past sexual history. For example, in Indiana, evidence of past sexual conduct of a sex crime victim is first considered by a trial judge, out of the presence of the jury, who determines if the evidence is or is not protected by the rape shield law.

4. This advice was given by the chief of police to my class of students as they began their research project on rape.

REFERENCES

Brewer, G. D., & deLeon, P. (1983). *Foundations of policy analysis.* Chicago: Dorsey Press.

Brownmiller, S. (1975). *Against our will: Men, women, and rape.* New York: Simon & Schuster.

Ferguson, K. E. (1984). *The feminist case against bureaucracy.* Philadelphia: Temple University Press.

Fain, M. M., Richardson, L., & Wemmerus, V. A. (1992). Feminist rape education: Does it work? *Gender & Society, 6,* 108-121.

Gibbs, N. (1991, June 3). The clamor on campus. *Time,* pp. 54-55.

Harding, S. (1991). *Whose science? Whose knowledge?* Ithaca, NY: Cornell University Press.

Herson, L.J.R., & Bolland, J. M. (1990). *The urban web: Politics, policy, and theory.* Chicago: Nelson-Hall.

Jenkins, M. J., & Dambrot, F. H. (1987). The attribution of date rape: Observer's attitudes and sexual experiences and the dating situation. *Journal of Applied Social Psychology, 17,* 875-895.

Kelly, L. (1988). *Surviving sexual violence.* Minneapolis: University of Minnesota Press.

Kemp, A. A. (1994). *Women's work: Degraded and devalued.* Englewood Cliffs, NJ: Prentice Hall.

Koss, M. P., Dinero, T. E., Seibel, C. A., & Cox, S. L. (1988). Stranger and acquaintance rape: Are there differences in the victim's experience? *Psychology of Women Quarterly, 12,* 1-24.

Krahe, B. (1988). Victim and observer characteristics as determinants of responsibility attributions to victims of rape. *Journal of Applied Social Psychology, 18,* 50-58.

Martin, P. Y., & Hummer, R. A. (1989). Fraternities and rape on campus. *Gender & Society, 3*, 457-473.

Pagelow, M. D. (1984). *Family violence*. New York: Praeger.

Reinharz, S. (1992). *Feminist methods in social research*. New York: Oxford University Press.

Roberts, C. (1989). *Women and rape*. New York: University Press.

Warshaw, R. (1988). *I never called it rape*. New York: Harper & Row.

11

Women's Need for Child Care: The Stumbling Block in the Transition From Welfare to Work

ALMA H. YOUNG
KRISTINE B. MIRANNE

In his new "Partnership for America" program, President Bill Clinton has included among the participants those individuals receiving public assistance, commonly known as welfare. The majority of these persons are women with young children. Although he promises to provide opportunities, the president also demands greater responsibilities in return from these women.

Mr. Clinton's interest in welfare reform is long-standing. As chair of the National Governor's Council, he argued for the transition from welfare to work and was one of the architects of the bipartisan Family Support Act of 1988. The result of this initiative is that entitlements to cash income have been replaced by cash relief conditional on work effort. Single parents receiving Aid to Families With Dependent Children (AFDC), and who have children aged 3 years and older (aged 1 year at state option), are required to participate in job training, education, job search, or job placement programs. Transitional support services, including child care, are to be provided for one year following AFDC ineligibility due to increased earnings. At the same time, the individual states are to broaden their efforts at collecting child support payments from absent parents.

Given the current focus on efficiency and reduction of public monies, it is not surprising that this legislation focused on education and training. Although participants will be provided transitional services, the emphasis continues to be on job and educational training as the answer to ameliorating poverty. The issue not adequately addressed, however, is that the lack of affordable child care is one of the major barriers to women's participation in the labor force (Bloom & Steen, 1990) and that it likely contributes to women's lower earnings as well. Without adequate child care services, women with young children are effectively excluded from most types of employment and training.

In addition, although the needs of the child are said to be at the heart of this legislation, this is the first time that the federal government has required welfare mothers with children under the age of 5 to enroll for education, training, or employment (Vann, 1991). It is apparent that the underlying philosophy of the Family Support Act is that the family, as opposed to the government, should provide the primary economic support for its children. Yet we fear that policymakers do not have a full appreciation of the complexity of the situation of women and work. To gain that appreciation, it is critical that the dynamics of women and poverty, of the economy and work, and of women's work in all of its facets be fully recognized.

This lack of understanding is evident by the fact that the United States is the only major industrialized nation without a national policy regarding child care. Despite the growth in women's labor force participation over the past few decades, child care services have not expanded to meet the increasing demand. We also have no official data indicating how many women are excluded from employment or whose employment is restricted to part-time because of the lack of child care. Yet the benefits to society that would occur if adequate child care services were available would include increased labor participation of low-income and single parents (thus reducing the public burden of welfare), reduced absenteeism and job stress for working parents, and increased job productivity for those businesses likely to employ a predominantly female workforce (Bloom & Steen, 1990).

We thus direct our attention to evaluating the goals and objectives of the child care component as outlined by the Family Support Act. We begin with a brief discussion of women's poverty and welfare policies to date. We then outline Project Independence, Louisiana's response to the federal mandate. We are particularly interested in how these policies will be implemented within the state's largest urban area, the City of New Orleans. Several policy recommendations are offered based on a survey of potential participants, focus group discussions, and interviews with social service administrators and children's rights interest groups.

■ Women, Welfare, and Poverty

Although it is recognized that the majority of families receiving AFDC benefits are headed by women, there is little understanding as to how women's poverty differs from that of men. Most of the academic and legislative debates tend to equate women's poverty with that of men—the assumption being that women share men's resources. This premise has been the basis for defining the social policy debate as well as for the formulation and implementation of policy directives (Kemp, in press). We argue that this framework has resulted in serious misconceptions about the causes of poverty, the strategies employed to address poverty, and the measurement of success of social welfare policies (see Gordon, 1992).

Various theories of poverty have been used as a justification for policy reform. These theories can be divided loosely into two groups: individual (e.g., "culture of poverty" thesis) and structural (e.g., the rise in female-headed households). At best, these explanations only describe women's poverty; worse, they either blame women for being poor (i.e., for not having a husband to support them and their children) or characterize their economic status as that of dependents. In either case, the argument seems to be that women's chances for moving out of poverty are tied to their chances of being attached to a man. The lack of jobs produces poverty for men, but the paucity of husbands is apparently the source of women's poverty (Scott, 1984).

We can trace our current complex rationalizations for poverty to the Social Security Act of 1935, an omnibus law considered the foundation of our welfare state. Welfare scholars usually divide the programs into two categories: social insurance, which is more generous and more popular; and public assistance, which is more stigmatizing and less generous. The two major forms of social insurance, Old Age Insurance and unemployment insurance, disproportionately serve White men. These programs are considered respectable, stipends are relatively high, and they are received as a matter of entitlement without means testing.

Public assistance, however, serves mainly women and children and is often considered to be pejorative. Payments are low; standard amounts keep recipients below the poverty line and are not designed to allow individuals to attain a decent standard of living. Welfare payments are carefully calculated to be less than the lowest paid jobs so that poor people in general will seek employment over public assistance, if at all possible, and women in particular will seek marriage (Caputo, 1989, p. 90).

Interpreting public assistance programs as antipoverty efforts ignores the basic premise of eligibility—eligibility requires continual destitution. AFDC is actually a consumptive transfer—benefits are used up each month, leaving recipients just as poor at the end of the month as they were at the beginning (Beechley, 1984). Thus welfare and antipoverty have not been synonymous, although the concepts, even today, would have the public believe they are.

Historically, child welfare programs assumed that most women were mothers, that children required their mother's exclusive care, and that men would support them both (see Miller, 1990). As the welfare rolls grew, and as increasing numbers of women entered the labor market, policymakers turned to measures designed to push single mothers into the workforce. Handler and Hasenfeld (1991) point out, "The respectability of working married mothers only heightened the perceived deviance and moral depravity of single mothers, especially those with children born out of wedlock, who are on welfare rather than working" (p. 137).

The necessity of women with children at all income levels to work changed the attitudes of many, including liberals who historically had opposed punitive welfare policies by arguing that poor women should have the same right as other women to stay home and care for their children (see Abramovitz, 1988).

Gender roles have also shifted dramatically during the past 35 years, particularly because of the rapid change over the past two decades in the nation's demographics as more women became heads of their families (see Pearce, 1978; Sarvasy & Van Allen, 1984). These changes suggest that policymakers should construct programs that would mitigate the economic disadvantages of divorce and being a single parent. Instead, there appears to be little understanding as to the determinants of women's poverty or how to address the problem. Women's increased vulnerability to poverty is not just a function of their inability to marry or stay married (as many would have us believe), but rather, a function of the discriminatory treatment of women in the labor market combined with the allocation to women of the responsibility of children and the lack of affordable child care (Caputo, 1989; Kemp, in press; McLanahan, Sorenson, & Watson, 1989).

In addition to the issue of gender, racial discrimination also becomes important as we discuss the poverty of women and their children. The high percentage of Black female-headed families may be a recent historical phenomenon, but their poverty is not (see Wacquant & Wilson, 1989).

Although it is particularly Black women and children who are poor, it is not simply because they are more likely to live in female-headed households. For example, two out of three Blacks living in female-headed households were already poor *before* a change put them into that family configuration (Bane, 1986).

The point being argued is that women who are receiving AFDC are among the poorest of the poor. These same individuals are likely to have limited education and work experience. This is particularly the case in those states that have traditionally been less than generous in providing welfare benefits. Poverty in Louisiana is severe when compared to other states. Louisiana ranks 51st in percentage of its children in poverty; children's poverty rates increased from 23.5% in 1979 to 34.6% in 1990. The state ranks 49th in percentage of children in single-parent families, which increased from 25.3% in 1984 to 31.3% in 1991 (*Kids Count Data Book*, 1992). In Louisiana, an AFDC family of three receives a monthly cash grant of $190 ($1,656 per year). This is $468 less per month than the $658 need standard established by the state, and $690 less than the federal poverty level for a family of three. In other words, a family of three on AFDC in Louisiana receives an entitlement that is 22% of the annual federal poverty level established for a family that size. It should be noted that although no state brings a family to 100% of the poverty line, the nation's median is 47.5% (Joseph, Gilbert, & Tuman, 1991).

As of 1990, the reported statistics for New Orleans, a city whose population was 61.9% African American, reflect a large proportion of urban poor within a poor state:

- 11.9% of the city's households (24,094) were receiving AFDC payments. African Americans composed 97% of this population.
- 23.8% of the city's households (48,348) were receiving food stamps. Black households represented 91% of this group.
- 27% of the children aged 19 years and younger (47,398) live in families that receive AFDC (Joseph et al., 1992).
- Citizens of New Orleans were 25% more dependent on social service safety net transfer payments than the average person in the United States (*Metrovision*, 1990).

In sum, the AFDC population in New Orleans can be characterized as being predominantly Black, with very low income, highly dependent on public assistance, and at high risk due to its persistent level of poverty.

■ **Project Independence:**
   Louisiana's Response to Welfare Reform

Project Independence is the name of the Louisiana Jobs Opportunity
and Basic Skills (JOBS) program mandated by the Family Support Act
of 1988. In May 1989, a statewide task force recommended that there
be six components of Project Independence: (a) education, (b) skills
training, (c) job readiness activities, (d) job development and job placement, (e) job search, and (f) on-the-job training. Case managers were to
be trained to assist participants in assessing resources and opportunities
required for self-support. In addition, the first year of implementation
would be limited to six parishes (counties), including Orleans Parish
(coterminous with New Orleans). The federal mandate exempts mothers
with children aged 3 years or younger, although states were given the
option to require participation from mothers with children 1 year or
older. Louisiana chose to exercise the latter alternative.

The New Orleans plan was developed by an Advisory Committee
composed of 41 persons representing a cross section of the community.
It included one welfare recipient and the head of the Welfare Rights
Organization. The committee's charge was to design the program components that would be responsive to the available resources at the local
level. The program was to be executed through a parishwide management system using identified community resources to accomplish the
goals and objectives of the program.

In preparation for the local implementation of the program, the Louisiana Department of Social Services conducted a survey in April 1990 of
10,942 potential mandatory participants (i.e., those custodial parents who
met the criteria requiring participation in the program). Groups targeted to
participate in Project Independence included recipients who had received
AFDC in 36 of the last 60 months, custodial parents under the age of 24
without a high school diploma and/or with no work history, and recipients
whose youngest child was age 16 years or older.

Table 11.1 reflects the age categories of the potential participants.
The majority of these individuals (68%) were between the ages of 20
and 34 years old. Although this age group may be considered prime
candidates for employment, they were also more likely to have younger
children that require either all-day care or before- and after-school care.
In comparison, only 28% of the respondents were age 35 years or older.
This age cohort may need only part-time child care if their children are
already enrolled in school.

TABLE 11.1 Age of Potential Participants

Age	Number	Percentage
Under 20 years	341	3
20-23 years	1,521	14
24-34 years	5,898	54
35-44 years	2,409	22
45 years and older	605	6
No response	168	1
Total	10,942	100%

SOURCE: Louisiana Department of Social Services (1990).

TABLE 11.2 Education Level of Potential Participants

Education Level	Number	Percentage
Grades 1-8	855	8
Grades 9-11	4,657	43
High school graduate	3,986	36
Some college	1,133	10
College graduate	101	1
No response	210	2
Total	10,942	100

SOURCE: Louisiana Department of Social Services (1990).

Table 11.2 outlines the education levels of the respondents. A majority, 51%, did not have a high school diploma, with 8% of that group reporting less than a ninth-grade education. High school graduates composed 36% of the total, and 11% had some college education. This indicates that a significant proportion of program resources would need to be targeted toward assisting participants to complete high school or attain the graduate equivalency degree (GED). Of note, the survey also revealed that 12% of the potential participants were already involved in self-initiated training or education. Of that group, 23% were working toward completing high school or the GED, 33% were pursuing vocational/technical training, and 26% were enrolled in a college or university.

Another question asked of the respondents referred to their ability to secure transportation. The vast majority of participants depend on public transit—a factor that is often considered only in light of its expense. Reimbursing individuals for the cost of transportation is an important support service; however, the time element involved must also be recognized. Women add several hours to their working day if they have to transport themselves to a place of employment or training while also taking their children to child care centers. Over 90% of the potential participants reported that they depended on public transportation in order to be mobile.

This survey, conducted prior to the implementation of Project Independence, highlights several issues. For women to be able to remove themselves from the welfare rolls, they must secure employment that will support their families. Yet the young age of these women and the low level of their education present a major challenge. The heavy dependence on public transit is also a concern, reflecting but one of the everyday logistical problems that these women with children must address to go to school or to work. For young women with children, there is also the need for adequate child care services.

■ Louisiana's Child Care System

Any analysis of poverty among women must take into consideration the fact that these women are not only responsible for rearing their children but are also overwhelmingly the primary financial caretakers of their children. This dual responsibility often leads to their secondary economic status, because they do not have the opportunity to build a consistent work history. The result is that women have difficulty finding employment that pays more than the prevailing minimum wage or that provides benefits such as health care. Although Louisiana has readily accepted the federal mandate of welfare reform, financial constraints limit its ability to make a full range of services available to those recipients seeking assistance. Yet to become economically sufficient, women must not only have jobs that pay a decent wage, they must have access to supportive services.

The states are mandated to guarantee child care "to the extent that it is determined by the agency to be necessary" for an individual's employment, training, or educational activities, and if the state determines that the individual is "participating satisfactorily" (Vann, 1991). One of the expected impediments to the realization of the legislation's goals, however, is the lack of commitment by the states to provide child care for all those who need it, and thus, their inability to condition mandatory participation because of the lack of child care (Miller, 1990). Critics of

TABLE 11.3 Number of Children Who Would Require Child Care, by Age

Ages of Children	Number	Percentage
0-1 year (infant)	1,611	8.0
2-3 years (toddler)	3,771	18.9
4-5 years (preschool)	3,592	18.0
6-12 years (school-age)	10,562	52.9
Youngest child 16 years and older	445	2.2
Total	19,981	100.0

SOURCE: Louisiana Department of Social Services (1990).

the legislation fear that states will not be able to assist in the creation of the large numbers of child care spaces that will be needed, will not be able to attract qualified contractors who meet state and federal standards, and will not be able to help contractors in attracting qualified child care workers. The result will be that the burden of finding child care will either shift to the recipient, with very little formal oversight, or that there will be a loosening of the compulsion to participate from those who fail to locate necessary arrangements (Sanger, 1990).

Although the focus of the Family Support Act is on the adult recipient, there is a sizable population of children also affected by this reform. The preliminary survey conducted in Orleans Parish collected data as to the number of children who would be included in Project Independence efforts. Given the ages of these children as depicted in Table 11.3, we would expect that approximately 45% would require full-time child care and almost 53% would need part-time care.

In developing its program, the State of Louisiana outlined two components of child care service. First, transitional child care was to be implemented by April 1990 to provide child care services to certain AFDC clients who become ineligible for a public assistance grant. To receive the benefits of this transitional service, a client must have received AFDC benefits for 3 of 6 months immediately preceding ineligibility and must have become ineligible because of earnings. The child care service would be available for up to one year following ineligibility for an AFDC grant. The intent was to assist individuals in "bridging the gap" between dependence on public assistance and economic independence. The client was required to contribute to the cost of this service on a sliding scale.

Second, JOBS-related child care was to be implemented concurrently with the JOBS program (i.e., child care expenses would be provided while the custodial parent was involved in education, training, or job search). Child care reimbursement was to be guaranteed to mothers who had dependent children under the age of 13 or for those children who were physically or mentally incapable of caring for themselves.

In addition to providing financial support for child care, the state encouraged recipients to select the child care services best suited for their needs. The Governor's Task Force (1989) stated further that "AFDC clients and former clients who qualify for these benefits . . . avail themselves, for the benefit of their children, of the highest quality child care available" (p. 34).

Although the call for the highest quality child care is a valiant statement, the reality is that the child care system in Louisiana/New Orleans is characterized by a low level of quality, fragmentation, an inadequate financial base, and a lack of skill on the part of many child care workers (interview with J. Watts, executive director, Agenda for Children, May 11, 1992). Child care costs tend to be low in Louisiana, perhaps reflecting the lower incomes of its workers. Low child care costs, however, result in the low wages of child care workers, which in turn lead to a large turnover in staff (approximately 50% per year). Louisiana also licenses two classes of child care providers: Class A and Class B. The former are considered to be formal child care centers and must meet physical and caretaker/pupil standards. The latter, informal child care centers, are facilities governed by much less stringent standards. These centers, for instance, are not required to carry liability insurance and allow for corporal punishment. Other types of informal child care facilities include family center care (six or fewer children in the home of the caregiver) and school-based care (which is only beginning to acknowledge the fact that there are working hours other than 9 a.m. to 3 p.m.).

Another issue that must be addressed is the actual availability of slots within formal child care centers. Before Project Independence, Louisiana had 3,618 children in subsidized care, with a waiting list of 10,590. In New Orleans, as of January 1990, there were 1,162 children in subsidized care, with a waiting list of 6,249 (interview with J. Watts, May 11, 1992). After reviewing the resources within the state's communities, the Governor's Task Force anticipated that most of the child care in Louisiana would be provided by informal facilities. Thus the task force recommended that the choice of child care be made by the JOBS participant, with assistance from her case manager who would have an automated listing of licensed child care providers. Family center child care would be an option available to the JOBS participant, and again the case manager would assist with this alternative. Efforts would also be made to identify child care providers for older children (ages 6 to 13 years), through the use of before- and after-school programs and summer camps. Although the state plan did not mention informal child care arrangements through relatives and friends, that, too, was an option through the federal legislation.

Access to a wide variety of child care arrangements is critical to the success of Project Independence. Before they can fully participate in this program, women must be assured that their children are receiving adequate and dependable care. The responsibility has been placed on the participant to locate this care. The long waiting lists for subsidized formal child care centers will force women to rely on informal networks. Past ethnographic research has indicated that Black single parents have been able to depend on a strong intergenerational system consisting of friends and family who are ready sources of social and economic support (Stack, 1974). Recent research has reported that, regardless of race, single mothers have better access to kin-based support than do married mothers. Many mothers, however, do not have this assumed support and, if they do, the support received is insufficient to provide access to adequate and sustained child care (Hogan, Hao, & Parish, 1990).

Welfare mothers also express the same concerns about child care as do other working mothers. Research by Sonenstein and Wolf (1991) reported that convenient hours and location, good adult supervision, low child-to-adult ratios, learning opportunities, and the child's happiness, were more important to AFDC mothers than whether the care provided was through formal or informal arrangements. For example, almost 30% of Sonenstein and Wolf's sample population required child care before 7:00 a.m. and after 6:00 p.m., hours not conducive to using formal child care centers.

The survey conducted in Orleans Parish also asked respondents which type of child care arrangements that they would prefer to have provided by the state. Table 11.4 reports that a variety of child care options are preferred. Almost 36% would select a formal, licensed child care center. Another 21% would prefer the state to provide unlicensed arrangements (e.g., family center care). However, almost 30% would request no help from the state in providing child care. This group likely included those who have a support system in place that provides child care. Therefore they opt for no intervention by the state. This group may

TABLE 11.4   Child Care Arrangements Preferred by Potential Participants

Child Care Arrangements	Number	Percentage
Type A (licensed child care center)	3,925	35.9
Type B (unlicensed arrangement)	2,301	21.0
Type C (either A or B)	1,156	10.6
Child care available at no cost	339	3.1
Self-provisioning	3,221	29.4
Total	10,942	100.0

SOURCE: Louisiana Department of Social Services (1990).

also include those respondents with children of an age where child care is no longer considered necessary.

It cannot be emphasized enough that welfare mothers should be able to choose the child care arrangements that are best suited for themselves and their children. To better understand how women choose child care services, we conducted two focus groups centered on the issue of who should provide these services. The participants were enrolled in a literacy program that targeted both the parent and child. We were interested not only in their current child care arrangements but also in their future plans once they had completed their educational training and would presumably be entering the workforce.

The majority of the individuals were using family members for child care. In some instances, the children were infants and were considered to be too young for formal care. Yet even those mothers with toddlers and older children still opted to place their children with family members. The overriding concern was that the children be safe, be fed nutritious meals, and have structured schedules that included play time as well as naps.

When asked what they would look for in a formal day care setting, the respondents listed several of the same needs reported by Sonenstein and Wolf (1991)—low adult-to-child ratios, learning opportunities, cleanliness, safety, and proper discipline. Mothers indicated that before placing their child in a formal setting, they would spend several hours at the center to observe how the children were treated. Of those women who had previously used formal day care arrangements, most seemed satisfied with their child(ren)'s experiences. There was some concern, however, about the expense of this type of care as well as the convenience of operating hours.

Although most of these women preferred informal arrangements, they were emphatic that these caregivers have some level of experience with children. For example, a sister who did not have children of her own was not considered an appropriate child care provider. Interestingly, good friends were not used as much as we had anticipated. Women would leave their children with a friend for short periods of time (running errands, going to the grocery store, etc.) and would often exchange baby-sitting favors with each other. Before they would consider these individuals for full-time care, mothers wanted to be sure that they had some sort of formal child care training.

In sum, welfare mothers are seeking the same quality of child care that working parents require. The lack of adequate and affordable child care is a major barrier that must be overcome as women remove themselves from welfare dependency. For this reason, it is critical that the child care component of the welfare legislation receive serious review.

■ What Have We Learned?

Project Independence became operational on October 1, 1990, in 10 parishes. We can now ascertain several facets of the child care component. First, of the states that experimented with workfare prior to the enactment of the Family Act, Louisiana ranked near the middle in the percentage of funds allocated for child care subsidies (35.4%). In the JOBS program nationwide, child care costs as a proportion of the total program ranged from 50% in New York to 20% in Mississippi (Sanger, 1990).

Second, in Louisiana, the maximum allowance for children over 2 years of age is $175 per month. For those children aged 2 years or younger, $200 per month is the maximum allowable benefit. Children requiring part-time child care (20 hours or less per week) are allowed no more than one half of the applicable rate. Initially, care provided by unlicensed care givers or family centers qualified for only one half the $175/$200-per-month rate, even if the child was in care for more than 20 hours per week. At present, there is a marginal difference between rates for licensed and unlicensed care for children under 2 years of age and no difference in the reimbursement rate for older children. The monthly amount allowed for child care does allow a mother in New Orleans to find child care, although most services received for this level of expenditure may be minimal in terms of quality (Sonenstein & Wolf,

1991). Higher-quality care (i.e., those centers that meet national standards) can cost as much as $75 to $80 or more per week, substantially higher than the child care allowance.

Third, child care vendors have already begun to complain about the transient nature of the child care arrangements (interview with J. Watts, May 11, 1992). The instability stems from absenteeism or the need for mothers to change the location of the child care center as they move from one training site to another, from training site to job, and so on. For a mother making child care arrangements, location is often as important a factor as cost.

Fourth, child care is difficult to find for infants, regardless of the ability to pay for care. A recent national study shows that less than 10% of the vacancies in child care centers were open to children under the age of 1 year (Agenda for Children, 1992, p. 6). Because mothers in Louisiana are being required to participate in the program as soon as their youngest child reaches the age of 1 year, the need for infant care is great.

Fifth, we anticipate that the majority of JOBS participants will be employed by the service industry. This reflects not only the national economy but the economy in New Orleans in particular. These types of jobs are often at night or during odd hours—times when traditional child care services are not available. In addition, the majority of the participants must depend on public transit to travel to their place of training or employment. Formal day care centers usually close by 6 p.m., resulting in many mothers not being able to use this type of care because it is not feasible.

Last, mothers are encouraged to turn to their case managers for assistance in finding child care, along with other support services. The majority of the participants in our focus groups were receiving AFDC. Although the state accepted this literacy program under the auspices of the JOBS component, these individuals's caseworkers had not informed the participants of the various support services available to them. Several women had their children in various day care arrangements; none were receiving a subsidy. We are concerned that the focus on the placement of individuals in education/training programs will be a full-time effort for caseworkers resulting in little attention to other elements of the program.

■ Issues to Address

As we know, the established measure of success of policies that attempt to move women from welfare to work will be the actual reduction in welfare caseloads. Officials assume that the increased availability of subsidized child care will be an incentive for more women to leave the welfare rolls. Yet providing monies for child care will not in itself ensure success. Those who design and implement policy must be conscious of the specific needs of poor women in cities who are receiving subsidized child care. To this end, we suggest that options in child care arrangements, training women as child care workers, and maintaining supportive resources are especially important.

A majority of the women who participated in the Louisiana pilot program used subsidized formal child care centers. We are concerned, however, that as more women are required to participate in similar programs, there will be a shortage of available slots within subsidized formal centers. Across the country, the wait for placement in subsidized formal child care will only grow larger as more women participate in the JOBS program. Mothers with younger children will have problems meeting the job participation mandate because formal child care for younger children is not readily available. Informal arrangements, where family centers, friends, or relatives are providers of child care, will become necessary. In fact, mothers often prefer this type of arrangement for their infants. Sonenstein and Wolf (1991) determined that many AFDC mothers use out-of-the-home relative care for preschoolers, in part because of the child-to-adult ratios and their children's happiness.

Increasing child care options provides more alternatives to mothers who under other circumstances may not be able to participate in mandated training and education programs or accept employment because child care is not available. It is anticipated that many of the jobs available to participants will be within the service sector. Many of these positions require shift work during a time of day when formal child care is not available. This concern is exacerbated by the transportation time added, because most participants will have to rely on public transit, especially in urban areas.

One way to increase the number of child care options is to encourage the training of child care workers. Child care workers can be trained for work in both formal and informal child care arrangements. In fact, women who are currently caring informally for the children of other poor women can be better trained for the work they already do. In conjunction, if informal child care arrangements are adequately funded and these arrangements are encouraged, then child care can actually create jobs for women who are required to work. Women will be able to place their children within their known network of friends and relatives while being assured that these individuals have received formal child care training.

An individual employed at the minimum wage will not be able to afford the true costs of child care. Therefore, once a participant is employed and has completed her transition from welfare to work, she is likely still to require subsidized child care and assistance in obtaining the services of other programs that subsidize child care. For example, Congress has recently instituted the Child Care and Development Block Grant program. This program provides child care vouchers for low- and moderate-income families. Families receive assistance on a sliding scale, covering between 10% and 100% of the total cost of child care, depending on family size and income. Many women could benefit from such a subsidy (for example, virtually all Project Independence participants would qualify). If women are to continue to be employed and not return to welfare dependency, then all supportive resources must be made available to them even after they have completed employment training programs.

■ Summary

Until the inception of Project Independence, the lack of subsidized child care in Louisiana had been a major concern. State bureaucratic leaders are pleased by the number of women who have already left the welfare system. If Louisiana or any other state continues to measure the success of welfare-to-work programs in this way, however, it must recognize that the provision of child care services is as important as the jobs/training component. If women are assured that their children are receiving adequate child care, they will be better able to commit themselves to education and training programs that can lead to reduced dependency on financial assistance. Women need to be free, however, to choose the child care arrangements they prefer. For poor women, especially Black women, access to kin networks has been a major source of child care support. These mothers often prefer that child care be provided within these networks. Policy must recognize the validity of these informal child care arrangements. Further, it is important to find ways to enhance the economic well-being of these caregivers, such as providing training and income for the work they do.

REFERENCES

Abramovitz, M. (1988, September 26). Why social welfare is a sham. *Nation, 221*, 224-225.

Agenda for children. (1992, Spring). *Agenda for children newsletter* [newsletter]. New Orleans, LA: Author.

Bane, M. J. (1986). Household composition and poverty. In S. Danziger & D. W. Weinberg (Eds.), *Fighting poverty: What works and what doesn't* (pp. 209-231). Cambridge, MA: Harvard University Press.

Beechley, L. (1984). Illusion and reality in the measurement of poverty. *Social Problems, 31*, 322-333.

Bloom, D. E., & Steen, T. P. (1990). The labor force implications of expanding the child care industry. *Population Research and Policy Review, 9*, 25-44.

Caputo, R. (1989). Limits of welfare reform. *Social Casework, 70*, 85-95.

Gordon, L. (1992). Social insurance and public assistance: The influence of gender in welfare thought in the United States, 1890-1935. *American Historical Review, 97*, 19-54.

Governor's Task Force. (1989). *Governor's task force on welfare reform: Final report and recommendations*. Baton Rouge, LA: Department of Social Services, Office of Family Support.

Handler, J. F., & Hasenfeld, Y. (1991). *The moral construction of poverty*. Newbury Park, CA: Sage.

Hogan, D. P., Hao, L., & Parish, W. L. (1990). Race, kin networks, and assistance to mother-headed families. *Social Forces, 68*, 797-812.

Joseph, M., Gilbert, G., & Tuman, D. (1991). Project Independence: A description and analysis of potential economic and social benefits to AFDC families, their children, and the economic development of New Orleans. In G. Gilbert (Ed.), *The state of Black New Orleans, 1991* (pp. 163-199). New Orleans, LA: Urban League of Greater New Orleans.

Kemp, A. (in press). Poverty and welfare for women: The alternative between being deprived or destitute. In J. Freeman (Ed.), *Women: A feminist perspective*. Palo Alto, CA: Mayfield.

Kids count data book. (1992). Washington, DC: Center for the Study of Social Policy.

Louisiana Department of Social Services. (1990). *Survey of mandatory JOBS participants in Orleans Parish*. Baton Rouge, LA: Office of Family Security.

McLanahan, S. S., Sorenson, A., & Watson, D. (1989). Sex differences in poverty, 1950-1980. *Signs, 15*, 102-122.

Metrovision: An economic development plan for metro New Orleans. (1990). New Orleans, LA: Greater New Orleans Chamber of Commerce.

Miller, D. C. (1990). *Women and social welfare: A feminist analysis*. New York: Praeger.

Pearce, D. (1978). The feminization of poverty: Women, work, and welfare. *Urban and Social Change Review, 11*, 28-36.

Sanger, M. B. (1990). The inherent contradiction of welfare reform. *Policy Studies Journal, 18*, 663-680.

Sarvasy, W., & Van Allen, J. (1984). Fighting the feminization of poverty: Socialist-feminist analysis and strategy. *Review of Radical Political Economics, 16*, 89-110.

Scott, H. (1984). *Working your way to the bottom: The feminization of poverty*. Boston: Pandora.

Sonenstein, F. L., & Wolf, D. A. (1991). Satisfaction with child care: Perspectives of welfare mothers. *Journal of Social Issues, 47,* 15-31.

Stack, C. (1974). *All our kin: Strategies for survival in a Black community.* New York: Harper & Row.

Vann, B. H. (1991, November). *Childcare and the welfare state: Who's minding the kids and what is the state's role in picking up the tab?* Paper presented at the annual meeting of the Social Science History Association, New Orleans, LA.

Wacquant, L.J.D., & Wilson, W. J. (1989). Cost of racial and class exclusion in the inner city. *Annals of the American Academy of Political and Social Science, 501,* 8-25.

12

Strategies for Serving Latina Battered Women

ANNA M. SANTIAGO
MERRY MORASH

Just 20 years ago, women who were abused by their partners had limited social and legal resources available to protect them from abuse. During the 1970s and 1980s, this situation changed rapidly, resulting, in part, from a shift in public attitudes regarding government intervention in family matters. Since then, a coalition of shelter workers, community-based organizations, legal service lawyers, and law enforcement personnel have worked at the local, state, and national levels to develop and implement services to battered women. By the late 1980s, 1,200 shelters were operating in the United States. Moreover, 47 states, Puerto Rico, the Virgin Islands, and the District of Columbia had enacted extensive legislation on domestic violence (see Lengyel, 1990; Lerman & Livingston, 1983).

Although services expanded, Schechter (1982) argued that until the 1980s, the concerns of minority women were essentially ignored in the development of domestic violence programs. The limited theoretical and empirical work conducted on minority victims suggests that domestic violence services are underused by Black and Latina women (e.g., Coley & Beckett, 1988a, 1988b; Ginorio & Reno, 1985; Gondolf,

AUTHORS' NOTE: This research was partially supported by a grant from the Inter-University Committee for Research on Contemporary Hispanic Issues of the Social Science Research Council whose support is gratefully acknowledged. Additional support for this project was provided by the Population Studies Center, University of Michigan, and the School of Criminal Justice, Michigan State University. The authors would like to thank our interviewing team for their assistance in data collection. The research assistance of John Burrow, Chiquita Collins, Alissa Friedman-Torres, Robin Haar, and Grace Loo is also noted with gratitude. Special thanks to Donna Eder for her helpful comments on an earlier draft of this manuscript.

Fisher, & McFerron, 1988; Ubarry & Santiago, 1989). At the same time, the empirical evidence suggests that Black and Latina women encounter a similar, or even higher, incidence of spouse abuse than do majority women (Gondolf et al., 1988; Lockhart, 1985; Straus & Smith, 1990).

However, compared with Anglo women, victims who are racial, ethnic, religious, or linguistic minorities are not only less likely to seek shelter, they are more likely to encounter difficulties with service providers (Davis, 1984). Generally, services are provided in settings outside of minority communities, making physical access a major concern. Domestic violence programs have tended to ignore cultural differences in values and norms regarding issues such as appropriate help-seeking behaviors and child-rearing practices (Coley & Beckett, 1988a; Ginorio & Reno, 1985; Marin, Marin, Padilla, & Rocha, 1983). The failure to respond to cultural differences can lead to the imposition of a universal set of values and practices, which not only displays a high degree of insensitivity to the particular needs of minority women but also provides an additional level of victimization when service providers respond without any understanding of cultural differences or fail to consider minority concerns at all.

Information on the problems faced by Latina battered women[1] is particularly limited. This lack of information is compounded by their underuse of institutionalized services (Marin et al., 1983; Rothman, Gonant, & Hnat, 1985; Schensul & Schensul, 1982). Explanations for the lower levels of service use by Latina battered women have focused on four areas: (a) cultural norms that emphasize the use of extended-family networks in times of crisis, (b) failure to recognize the diversity of the Latino population, (c) limited access to existing services, and (d) the limited legal protection available to immigrant women.

Rothman et al. (1985, pp. 201-203) argue that regardless of the availability of external resources, Latinas would be reluctant to seek professional services because of their strong emphasis on familism. Cultural norms stress that problems are best addressed within the extended-family unit. Further, the discussion of family matters or personal feelings outside of the family unit is particularly discouraged. As a result, help from outside agencies is rarely sought (Zambrano, 1985).

This reluctance to use formal support services is compounded by perceptions that culturally sensitive resources are not available. The first shelters serving ethnic communities appeared in the early 1980s. It has been argued that in multi-ethnic communities, services need to be designed to respond to the needs of specific groups. With nearly two dozen distinctive nationality groups in the United States, responding to the needs of Latinos presents a tremendous challenge (Ginorio & Reno, 1985; Torres, 1991). Each group has had different social, economic, and educational trajectories that not only condition the types of family conflicts that may arise but also affect the service needs of this population. Often, shelters and other services for battered women do not respond to the unique needs of any group, let alone the needs of specific Latino groups. In the face of limited resources, it is highly unlikely that shelters would be able to even partially address these concerns.

Access to services is also constrained by the lack of bilingual, bicultural service providers. Relatively few shelters have the capability of serving non-English-speaking clients. In police departments and the courts, Latinas can encounter blatant hostility from service providers who have little sympathy for individuals with limited English ability (Ubarry & Santiago, 1989). This type of hostility compounds the difficulty of even those who are proficient in English in using the criminal justice system to protect themselves and other family members (Dobash & Dobash, 1992; Hirschel, Hutchinson, Dean, & Mills, 1992).

A fourth constraint affecting Latina victims is the limited resources and legal protection available to immigrant (particularly undocumented) battered women. Undocumented status precludes specific program participation (i.e., welfare benefits) that enable women to leave abusive relationships. Further, until recently, undocumented wives had extremely limited legal protection, which, in effect, placed them in a position of virtual enslavement to abusive partners who threatened to initiate deportation activities (Ginorio & Reno, 1985; National Woman Abuse Prevention Project, 1988; Zambrano, 1985).

To our knowledge, only one study has been conducted to date that empirically assesses the types of services as well as service use patterns of Latina victims of domestic violence. In a study of 169 agencies that provided services to victims of domestic violence, Ubarry and Santiago (1989) examined patterns of agency exposure to Latina victims, the types of programs and staff available to meet the needs of Latina victims, and provider perceptions of the service needs of Latina victims.[2] This study reveals that in a state that had a Latino population of nearly one half million, approximately one third of the agencies providing services to victims of domestic violence had not served any Latinas during the period of study. Among those agencies that had served Latinas, half of them had served fewer than 10 Latina clients. Only one fifth of the agencies responding to the survey reported serving 30 or more Latina clients. Those agencies most likely to serve Latina clients were law enforcement and medical facilities, indicating that, generally, only the most serious cases were identified within the system.

Of particular interest in regard to questions about factors affecting service use, Ubarry and Santiago (1989) found that relatively few Latinos were employed in any of the agencies (with the exception of Latino community-based organizations). One quarter of all agencies did not have any bilingual staff at the time of the survey. Moreover, another 22% had only one bilingual staff member. The lack of bilingual staff has been cited as one of the most salient factors affecting the use of formal services by Latinos (see Rothman et al., 1985; Zambrano, 1985).

Each respondent was asked to identify what was perceived to be the unmet service needs of Latina victims (see Ubarry & Santiago, 1989, pp. 34-49). Respondents reported that services to Latinas would improve through expanding outreach to the Latino community, increasing the number of bilingual/bicultural staff, developing community education programs, creating support groups specifically targeted for Latina victims, and implementing cultural sensitivity training programs for staff members. The insensitivity of staff, the inaccessibility of domestic violence programs, and the lack of contact with the Latino community were also cited as factors contributing to the low level of service use by Latina victims.

The Ubarry and Santiago (1989) study does not consider one important dimension affecting service use: client knowledge of and perceptions about existing domestic violence programs. The present study examines this dimension with data from a 1991-92 study of Latino family life. Specifically, we focus on the use of existing social services by Latina women who are victims of spouse abuse. Further, we explore the use of Latino community-based organizations as well as informal support networks as alternative strategies adopted by battered Latinas to deal with abuse by their partners.

◼ Data and Subjects

Data for this analysis were obtained from a 1991-92 survey of Latino family life conducted in a large Midwestern metropolitan area. Only Latina women who were 18 years old or older and married or cohabiting with a Latino male partner at the time of the survey were eligible to participate. A respondent could enter the study in two ways. First, her household could have appeared on a sampling frame prepared by Santiago for an earlier study of Latino health. This frame included telephone listings that had been updated and supplemented with household listings from local community agencies. Second, a woman could enter the study if a participant identified her as eligible to participate. We specifically asked respondents to identify women who either had an unlisted phone or did not have a telephone. These efforts yielded a total sample of 176 women. It is important to note that our aim was to collect information from a more representative sample of Latina women than previous studies, most of which have relied on small, clinical populations.

Each respondent completed a face-to-face interview with a bilingual interviewer at a place where the respondent felt comfortable and safe. Interviews were conducted in the respondent's language of preference. The average interview lasted 1.5 hours. Respondents were asked to complete a detailed questionnaire about family life, which included specific modules of questions on topics such as social support networks, documentation of acts of physical and psychological abuse, help-seeking behaviors, and contact with formal social service networks. The survey instrument combined the use of closed-ended questions with a series of open-ended questions that enabled respondents to elaborate on their experiences. In this chapter, we draw on both qualitative and quantitative data and present initial findings regarding the help-seeking strategies considered and employed by Latina women experiencing physical and psychological abuse from their partners.

Our sample consisted primarily of women of Mexican descent (80%). The median age was 37 years, and the average family size was 4.5 persons. Approximately 75% of the women were in their first marriage. Six out of 10 respondents had not graduated from high school. Slightly less than one third of the women were employed at the time of interview. Approximately half of the families had earnings in 1991 of $20,000 or less. Although respondents had lived in the metropolitan area an average of 11 years at the time of the interview, nearly one quarter of the women in the sample had lived in the area for less than 5 years. Although most women (53%) resided in predominantly non-Latino neighborhoods, friendship networks remained predominantly Latino. Nearly three quarters of the respondents reported that Spanish was the principal language used at home. Nine out of 10 of the women indicated that they were Roman Catholic.

■ Incidence of Abuse—Identifying the Potential Need for Service Delivery

When assessing service use and service delivery needs, it is important to document the number of women who experienced some level of physical or emotional abuse. Straus and Smith (1990) suggest that one out of every four Latino couples experience physical violence in any given year. Sorenson and Telles (1991) document that one fifth of foreign-born Mexican and nearly one third of Mexican American respondents reported violence at the level of hitting or throwing things at a spouse. Our findings suggest that Latinas experience fairly high levels of physical or emotional abuse. Four out of 10 respondents reported experiencing physical abuse during the 12 months prior to the interview. Seven out of 10 women indicated that they had been emotionally abused by their partners. Over one half of the women indicated that they had experienced both physical and emotional abuse.

Among women reporting physical abuse,[3] the most common acts of physical violence included acts that their partners did to spite them (33%), being slapped by partner (16%), being pushed by partner (14%), being hit with something by their partner (10%), and being forced to have sexual relations (10%). Among the women reporting emotional abuse,[4] the most frequently cited acts included being told that their partners knew the "right" way (55%), having television or other activities being more important than respondent (51%), having to second-guess partner (42%), being worried about his reactions (40%), coping with partners' jealousy (30%), coping with partner's criticism (29%) or ridicule of respondent's family (26%), or having money withheld from respondent (23%).

Thus we can see that a relatively high fraction of Latinas experience difficulties with an abusive partner. Generally, reports of emotional abuse were much higher and occurred more frequently than acts of physical abuse. The degree to which these acts are seen as abusive as well as the extent to which women perceive that seeking help in response to these acts is appropriate will shape patterns of service use. In this chapter, we focus on the latter issue—appropriate help-seeking behaviors—to better understand patterns of service use among Latina battered women.

■ Appropriate Sources of Support

As has been reported in earlier studies (i.e., Jacques, 1981; O'Connor, 1990; Rothman et al., 1985; Schensul & Schensul, 1982), Latinas

TABLE 12.1 Appropriate Sources of Support for Conflicts With Partners
($N = 173$)

	N	Percentage of Total
Informal support networks	126	72.8
Parents	32	18.3
Siblings	25	14.3
Children	25	14.3
Other family members	17	9.7
In-laws	17	9.7
Formal support networks	21	12.1
Clergy	18	10.4
Counselors	2	1.1
Other agencies	1	.6
Resolve between partners	17	9.7
No outside contacts	9	5.1

SOURCE: Compilations by authors from 1991-92 Latino Family Life Project Data.

encountered strong cultural norms to use informal support networks in times of family crisis. Since then, little research has examined how these support networks function in the case of spouse abuse. Therefore, all respondents in this study were asked to identify the person or persons to whom they would turn for support if they were experiencing some type of conflict with their partners. From their open-ended responses to this question, we identified who respondents considered to be appropriate sources of support (see Table 12.1).

As anticipated, approximately 73% of the women reported that if they were having problems with their partners, they would seek the support of family members. Among respondents seeking help from family members, 25% would talk to their parents, 20% would contact siblings, and another 20% would obtain help from their children. Other contacts within the family network included respondents' extended kin and the parents or extended kin of their partners.

Only 12% of the respondents indicated that they would seek the support of persons within formal institutional settings. Among this group, the majority of the women (86%) indicated that they would seek the services of the clergy for help in resolving marital conflicts. This supports previous research by Zambrano (1985) and Chalfant et al. (1990) that suggests that counseling might be sought from the clergy or other persons of trust. Only three women in this group reported that they would seek help from a counselor or other agency personnel. Thus it appears that Latina women are embedded in a social network where problems are discussed and resolved within the larger Latino community—specifically the extended family—before seeking help from formal institutions. Further, when help was sought from outside the family network, it would be most likely from the clergy.

About 10% of the respondents reported that they would resolve conflicts with their partners without outside intervention. As one respondent remarked, *"Los problemas con mi esposo son problemas de él y yo, nadie mas* [Problems with my spouse are problems between him and me, no one else]." Another 5% stated that not only would they not tell anyone about the conflicts, it was unlikely that there would be further discussion about the problems. This suggests that nearly one out of six women would have serious reservations about involving anyone in mediating conflicts with their partners.

■ Contacts With Existing Formal and Informal Networks of Support

Although it is clear that the majority of the women would seek the assistance of families and friends, it is important to examine the extent to which this actually occurs and the perceived effectiveness of these interventions. As Baca Zinn (1979, p. 65) suggests, within Mexican American families "cultural ideals and actual family behaviors are separate dimensions of family life." Despite the rather high levels of abuse, only 23% of the women reporting either physical or emotional abuse actually sought help to alleviate the situation at some point in the relationship. Most of this help seeking occurred within the previous 12 months before the interview, suggesting that there is some underreporting of help seeking for earlier periods.[5] Of interest, relatively few respondents (6%) cited any factors that would impede their ability to seek assistance, although a frequently cited problem was the partner's failure to cooperate.

In response to a question asking the women to identify sources of assistance, the most frequently used resources by women experiencing abuse were extended family members (54%), social service agency personnel (47%), and the clergy (36%). These findings should be interpreted with caution because the sample sizes are quite small and respondents could use more than one source of support. As we anticipated, informal family support systems were used frequently by respondents. However, the findings suggest the need to consider the role of formal support networks (i.e., counselors and clergy) as mediators in cases of domestic violence within the Latino community. Although it may be preferable to deal with abusive relationships within the extended-family network, our findings suggest that the involvement of trusted individuals, albeit outsiders, is not totally unacceptable. They also point to the need to adequately prepare service providers, particularly the clergy, to deal with these cases. Previous research (Chalfant et al., 1990; O'Connor, 1990) documents that a significant proportion of those of Mexican descent view the clergy and community-based organizations as important mental health resources. A number of respondents in the sample referred directly to specific Latino social service agencies based within the community as their primary sources for these services.

Latinas in our sample did not participate in the local domestic violence programs, nor did they frequently seek legal or medical assistance. Previous research suggests that local shelter programs may be perceived as inaccessible to the Latino community (Ginorio & Reno, 1985; Torres, 1991; Ubarry & Santiago, 1989). In the community we are studying, domestic violence programs for women of color have been primarily targeted to African American women. As a result, existing services are located in the African American community. Not only are these services located some distance from the center of the Latino community, they are also culturally distant.

Further, legal and medical institutions are looked on with distrust, to be used primarily in the most serious cases. Among our respondents, only 8 women have filed charges with the police, one because the man also threatened to kidnap their child. Two women reported that they had to seek medical attention because of the physical abuse, including one incident that sent both the respondent and her partner to the doctor.

When services were used, did respondents consider them to be effective? Respondents who received help from formal networks had mixed views regarding their effectiveness. Among those who sought help from the clergy, 50% reported that this help was effective. Of those seeking help from psychologists, family counselors, or other social service providers, 50% indicated that these services were effective. Nevertheless, a sizable proportion of respondents reported that these services were not effective. As one respondent commented, "*Recibimos orientación, aunque no nos sirvió para nada* [We received counseling, although it did not do anything for us]." Although our respondents did

TABLE 12.2 Expected Types of Assistance From Support Networks ($N = 150$)[a]

	N	Percentage of Total
Emotional or moral support	40	26.7
Anything respondent needed	31	17.9
Advice	29	19.3
Financial support	15	10.0
Shelter	12	6.9
Talk with partner	9	5.2
Other	14	8.1

SOURCE: Compilations by authors from 1991-92 Latino Family Life Project Data.
a. Excludes 26 persons with missing information.

not fully elaborate on what they believed to be the reasons for the limited effectiveness of counseling, we can draw from the research on Black battered women, which suggests that women of color come into the process with different expectations about the objectives of counseling. They also may have very distinct perceptions about the meaning of abuse (i.e., as another example of the oppression encountered by people of color). Further, effectiveness may be dampened by the lack of support from other family members for help seeking beyond the family unit (Coley & Beckett, 1988a, p. 489).

◼ Informal Support Networks as Alternative Help-Seeking Strategies

What were the kinds of support that women expected to receive from informal family networks? As shown in Table 12.2, 27% of all respondents reported that they would expect to receive *apoyo* (emotional or moral support) from their families. *Apoyo* really reflects the unconditional support of family members, especially in times of crisis. Another 19% indicated that their families would give them *consejos* or advice. This advice ranged from how to cope with the abuse to encouragement to leave the relationship. Smaller proportions of the respondents stated that their extended family would provide financial assistance (10%) or shelter (8%) if necessary. Of the respondents, 5% indicated that their families would talk with the partner to get him to modify his behavior.

Despite the fact that all respondents reported expecting to receive assistance from family members, nearly half of the respondents experi-

TABLE 12.3 Reaction of Respondents' and Partners' Extended Family Toward Physical and Emotional Abuse

	Respondents' Family				Partners' Family			
	Reaction Toward Physical Abuse (n = 41)		Reaction Toward Emotional Abuse (n = 48)		Reaction Toward Physical Abuse (n = 34)		Reaction Toward Emotional Abuse (n = 50)	
	N	Percentage of Total	N	Percentage of Total	N	Percentage of Total	N	Percentage of Total
Family did not know	19	46.3	23	47.9	11	32.4	13	26.0
Family did nothing	5	12.2	9	18.8	5	14.7	13	26.0
Told her to leave partner	8	19.5	2	4.2	1	2.9	—	—
Told him to leave	—	—	—	—	—	—	3	6.0
Calmed respondent/ partner down	2	4.9	1	2.1	0	0.0	0	0.0
Talked to partner	1	2.4	1	2.1	2	5.9	2	4.0
Blamed her	—	—	—	—	5	14.7	3	6.0
Other	6	14.6	12	25.0	7	20.6	16	32.0

SOURCE: Compilations by authors from 1991-92 Latino Family Life Project Data.

encing physical or emotional abuse did not tell family members about the abuse (see Table 12.3). A sizable fraction of the partners' families also were unaware of the abuse. In addition, 12% of the respondents who had been physically abused by their partners reported that, when informed of the incidents, their families did not get involved. For victims of verbal abuse, 19% of the respondents stated that their families ignored the situation. As might be expected, higher proportions of the partners' extended families ignored the abuse.

When families were involved, the respondents' families were more likely to encourage her to leave her partner. Indeed, family members would become angry if the respondent failed to act on this advice. As one woman stated, "*Mis padres se enojaban mucho porque no lo dejaba, pero nunca teniá el valor suficiente* [My parents got very angry because I would not leave him, but I never had sufficient courage]." Other family reactions to the physical abuse included telling the respondent to fight back, questioning why they were married in the first place, and providing shelter. Of interest, familial response to verbal abuse encouraged resignation to the situation. This attitude is succinctly stated as follows: "Stay with him, he is just having problems." Also, in-laws were much more likely to blame the respondent for her abuse.

Thus, despite respondent convictions of strong family support, we find that a sizable fraction of Latina women might not report abusive situations to their families. Further, in the cases where they do, the response may not be what is anticipated. One in 10 families ignored incidents of physical abuse, and 1 in 5 families chose to stay out of conflicts involving emotional abuse. Also once involved, these informal support networks may actually aggravate stresses both within the relationship as well as within the extended-family network. Therefore, although people in informal support networks may potentially serve as counselors/ mediators in marital conflicts. It cannot be assumed that this will always occur. As a result, a sizable number of victims may experience a high degree of social isolation because they feel uncomfortable or have been made to feel uncomfortable about revealing abusive situations.

Despite the preceding examples that underscore some of the negative reactions of family members, the degree of satisfaction for informal sources of support provided by family members mirrored that reported for services received from the formal sector. What does this mean? One can speculate that at least according to the perceptions of the respondents, both systems seem to be equally effective in dealing with spouse abuse. On the other hand, perhaps respondents have, in fact, lowered their expectations so that the response from their family may be better than what is anticipated from outsiders.

The question then arises, that if this is so, why seek the services of people outside the informal family network when the results are basically the same and, culturally, it is more appropriate to seek help from within? Or to put it another way, why seek the ostracism of the community for what is perceived to be limited assistance from outsiders when you will have to return to this community? One plausible response to these questions is that neither support system, functioning independently, adequately serves the needs of Latina women. As a result, a high percentage of Latina women in abusive relationships receive little or no support from either informal or formal support networks. Yet we can and need to speculate about how the effective use of and linkages between these networks would foster their use by battered Latinas.

■ Implications for Social Service Delivery Systems

Our findings confirm that there is very limited use of social services by battered Latinas. Relatively few Latinas seek help for spouse abuse in general, and when they do, these services are not procured from domestic violence programs. In the metropolitan area we studied, the severe shortage of shelter programs and the unavailability of bilingual personnel may effectively restrict service use. The lack of knowledge about existing programs may further limit their use by Latinas. Without outreach into the Latino community, battered Latinas may simply be unaware of the services available to them. As a result, both community leaders and service providers may be unaware of the need for services within the Latino community. In addition, limited contact with Latino communities, as with other ethnic communities, has led to their exclusion in the development, planning, and implementation of shelter policies and services.

Another factor affecting service use is that shelters and other services are located in areas that are far removed from the Latino community and that Latinas themselves are scattered among neighborhoods that are not predominantly Latino. This heightens the mistrust that Latinos already have of formal institutions. Also, because these services basically serve a non-Latina clientele, they may be perceived as being exclusively for Anglo or Black women. Although these restrictions may not exist, these images will persist as long as contact with the Latino community is limited and programs are developed without Latino involvement.

Further, Latinas need to feel that existing programs are responsive to their needs. As Coley and Beckett (1988a) reported was the case for Black women, shelter programs for Latinas need to be aware of the nuances of Latino culture and family life that would enable a battered woman to feel safe and comfortable. This again refers to the need for staff to be sensitive to specific needs of Latina women in terms of policy formulation and program development. In part, this means the creation of services that acknowledge how differences in language, culture, and tradition shape the response to the abuse. This leads to the development of various strategies to deal with the abuse and empowers women to actively make decisions about their own lives that are not judged in Anglo or class terms.

Moreover, it points to the need for community-based alternatives in which the strengths of the informal support networks are used. Rothman et al. (1985, pp. 201-203) stress that members of the extended family should be incorporated directly into the counseling process. They suggest that counselors should use family members as resources and advocates and take a family systems approach to resolving family problems. Coley and Beckett (1988a) suggest that an effective community-based alternative would be to develop battered women's self-help groups. When members of the extended family or the larger community are used as resources, Schensul and Schensul (1982) and O'Connor (1990) suggest that the presence of kin may actually facilitate the use of formal services. In part, this stems from the demystification of the counseling process as well as from broader community acceptance of these services as appropriate.

Although Latinas rarely use domestic violence programs in this community, it is important to note that they do seek help from community-based organizations such as churches and Latino-run agencies. Thus it appears that a critical factor shaping service use may be the availability of bilingual/bicultural services located within the Latino community. Even though relatively few respondents used formal services, those that did, reported that they were satisfied with the effectiveness of those services. Policymakers and service providers need to actively network with Latino community-based organizations to get the information out into the community, to serve as liaisons in the referral and counseling process, and to directly incorporate these service providers into the network through the development of jointly funded and operated programs.

It should be noted that the involvement of family and close friendship networks in working toward the reduction of abuse of women is complicated by the possibility that these networks will not support the woman in insisting on a stop to the violence. Although some of our data document positive support and intervention, this is not always the case. Similarly, one of the explanations for women remaining in abusive situations is that cultural or religious values support tolerance of abuse to maintain the family unit. From previous studies of Anglo samples (Gwartney-Gibbs, Stockard, & Bohmer, 1987; Smith, 1991), peer support of the husband for example, encourages abuse. Values and attitudes that are found in many different U.S. ethnic groups can support male dominance within the family and justify the use of violence to resolve conflicts between partners. Thus not only cultural sensitivity but strategies for producing fundamental changes in batterers' viewpoints need to be considered in designing and delivering services.

Finally, it should be emphasized that our formal institutions contribute to the acceptance of spouse abuse in nearly all societies. Thus we cannot exclusively target Latino communities and identify what is "pathological" about these communities that elicits such high levels of spousal violence. Rather, we need to look within the larger U.S. and global contexts and critically assess the factors associated with the escalation of wife abuse (Campbell, 1992). Only then will we be able to identify appropriate measures to serve victims and bring sanctions against perpetrators.

NOTES

1. In this study, the definition of *Latina* corresponds to the official census definition of Spanish origin. This definition includes persons of Mexican, Puerto Rican, Cuban, Central American, or South American descent. Furthermore, *battering* is defined to encompass physical, psychological, and sexual abuse.

2. In 1988, a mail survey was administered to agency heads (or their designated staff person) from 290 service agencies in New Jersey that provided services to victims of domestic violence. The sample included all of the domestic violence programs in the state, Hispanic community-based organizations, prosecutors's offices and Victim/Witness Assistance programs, family court judges, and all police departments, hospital emergency rooms, and alcohol treatment centers that provided specialized services to domestic violence victims or that were located in communities with Latino populations of 2,000 or more. The overall response rate was 63%.

3. Physical abuse was measured using Straus's Conflict Tactics Scale, which consists of 20 items, ranging from threats of physical violence to actual acts of violence, such as slapping, hitting, beating, use of weapons, and forced sexual relations. Each respondent was asked whether her partner had committed any of these acts during the 12 months prior to the interview or if it had ever happened during the relationship and to identify the number of times a particular act occurred during this period.

4. Emotional abuse was measured using the 55-item Profile of Psychological Abuse developed by Sackett (1992). This scale enumerates the type and frequency of acts such as manipulation, threats, and ridicule.

5. These estimates were based on answers to the following question: "As a result of the conflicts between you and your partner, have either one of you felt sufficiently bad or were sufficiently hurt to look for help from others?"

REFERENCES

Baca Zinn, M. (1979). Chicano family research: Conceptual distortions and alternative directions. *Journal of Ethnic Studies, 7,* 59-71.

Campbell, J. C. (1992). Wife battering: Cultural contexts versus Western social sciences. In D. Ayers Counts, J. K. Brown, & J. C. Campbell (Eds.), *Sanctions and sanctuary: Cultural perspectives on the beating of wives* (pp. 229-249). Boulder: Westview.

Chalfant, H. P., Heller, P. L., Roberts, A., Briones, D., Aguirre-Hochbaum, S., & Farr, W. (1990). The clergy as a resource for those encountering psychological distress. *Review of Religious Research, 31,* 305-313.

Coley, S. M., & Beckett, J. O. (1988a). Black battered women: Practice issues. *Social Casework, 69,* 483-490.

Coley, S. M., & Beckett, J. O. (1988b). Black battered women: A review of empirical literature. *Journal of Counseling and Development, 66*, 266-270.

Davis, L. (1984). Beliefs of service providers about abused women and abusing men. *Social Work, 29*, 244-250.

Dobash, R. E., & Dobash, R. P. (1992). *Women, violence and social change.* New York: Routledge.

Ginorio, A., & Reno, J. (1985, February). Violence in the lives of Latina women. *Working Together: Newsletter of the Center for the Prevention of Sexual and Domestic Violence*, pp. 13-15.

Gondolf, E. W., Fisher, E., & McFerron, J. R. (1988). Racial differences among shelter residents: A comparison of Anglo, Black, and Hispanic battered. *Journal of Family Violence, 3*, 39-51.

Gwartney-Gibbs, P. A., Stockard, J., & Bohmer, S. (1987). Learning courtship aggression: The influence of parents, peers and personal experiences. *Family Relations, 36*, 276-282.

Hirschel, J. D., Hutchinson, I. W., Dean, C. W., & Mills, A. (1992). Review essay on the law enforcement response to spouse abuse: Past, present and future. *Justice Quarterly, 9*, 247-283.

Jacques, K. (1981). *Perceptions and coping behaviors of Anglo-American and Mexican immigrant battered women: A comparative study.* Unpublished doctoral dissertation, U.S. International University, San Diego, CA.

Lengyel, L. B. (1990). Survey of state domestic violence legislation. *Legal Reference Services Quarterly, 10*, 59-82.

Lerman, L. G., & Livingston, F. (1983). State legislation on domestic violence. *Response, 6*, 1-28.

Lockhart, L. L. (1985). Methodological issues in comparative racial analyses: The case of wife abuse. *Social Work Research and Abstracts, 21*, 35-41.

Marin, B. V., Marin, G., Padilla, A. M., & Rocha, C. (1983). Utilization of traditional and non-traditional sources of health care among Hispanics. *Hispanic Journal of Behavioral Sciences, 5*, 65-80.

National Woman Abuse Prevention Project. (1988, August). Special issues facing the undocumented battered woman. *The Exchange*, pp. 10-16.

O'Connor, M. I. (1990). Women's networks and the social needs of Mexican immigrants. *Urban Anthropology, 19*, 81-98.

Rothman, J., Gonant, L. M., & Hnat, S. A. (1985). Mexican-American family culture. *Social Service Review, 59*, 197-215.

Sackett, L. A. (1992). *Assessing psychological abuse among battered women.* Unpublished doctoral dissertation, University of Michigan.

Schechter, S. (1982). *Women and male violence: The visions and struggles of the battered women's movement.* Boston: South End Press.

Schensul, S. L., & Schensul, J. J. (1982). Helping resource use in a Puerto Rican community. *Urban Anthropology, 11*, 59-79.

Smith, M. D. (1991). Male peer support of wife abuse: An exploratory study. *Journal of Interpersonal Violence, 6*, 512-519.

Sorenson, S. B., & Telles, C. (1991). Self-reports of spousal violence in a Mexican-American and non-Hispanic White population. *Violence and Victims, 6*, 3-16.

Straus, M. A., & Smith, C. (1990). Violence in Hispanic families in the United States: Incidence rates and structural interpretations. In M. A. Straus & J. Gelles (Eds.),

Physical violence in American families (pp. 341-367). New Brunswick, NJ: Transaction Books.

Torres, S. (1991). A comparison of wife abuse between two cultures: Perceptions, attitudes, nature and extent. *Issues in Mental Health Nursing, 12,* 113-131.

Ubarry, G., & Santiago, A. M. (1989). *Services to Hispanic victims of domestic violence: Provider perceptions of programs in New Jersey.* Final report prepared for the New Jersey Division on Women, Trenton.

Zambrano, M. (1985). *Mejor sola que mal acompañada: Para la mujer golpeada.* Seattle: Seal Press.

13

Gender, Race, and the Spatial Context of Women's Employment

IBIPO JOHNSTON-ANUMONWO
SARA McLAFFERTY
VALERIE PRESTON

As women's earnings and labor force participation have increased in Canada and the United States, more attention is being given to the spatial dimensions of women's employment. Such key issues as where women work and how their employment decisions are influenced by household responsibilities, labor market segmentation, and access to transportation are now being examined. A large body of research indicates that women work closer to home than do men and have shorter commuting times. This is one of the most consistent findings in the literature on women's spatial behavior. However, the gender differential in commuting is not ubiquitous but depends significantly on the social and spatial contexts in which women live and work. In particular, minority women often have long work trips, trips equal to or exceeding those of their male counterparts. This chapter explores the geographical dimensions of employment for minority women in the United States and Canada, using published and unpublished information drawn from various cities. Our review concentrates on the experiences of Black women, the largest group of minority women in the United States and a growing segment of Canada's minority population. We argue that any effective analysis of women's employment must recognize the diversity of women's experiences by considering how broader social divisions, such as race and ethnicity, shape women's access to jobs and income.

AUTHORS' NOTE: This research was supported in part by the National Science Foundation (Grant No. SES 9012916). We would like to thank the editors for their helpful comments on an earlier draft.

In both the United States and Canada, Black women's economic status differs substantially from that of White women. On average, African American women earn less and are more likely to be unemployed than are White women (Amott & Matthaei, 1991). The differences are less striking in Canada than in the United States, reflecting the smaller size of Canada's Afro-Canadian population and its very different history. The fact that a majority of Black women in Canada are recent immigrants also affects the economic gap there. Yet, in both countries, Black women are disadvantaged in the labor market (Amott & Matthaei, 1991; Statistics Canada, 1990). Also, a higher percentage of African American women than White women maintain their own households, with no husband present. Not only do these women bear full domestic responsibilities but also many work outside the home to support their families. Although the increase in female-headed households among African Americans has attracted attention, Brewer (1988) observes that

> an emphasis on female-headed households misses an essential truth about black women's poverty: Black women are also poor in households with male heads. With or without a male present, there is a strong likelihood that black women and children will be living in poverty in America today. (p. 334)

Thus the central issue in the well-being of African American women is not family structure but access to jobs and income (McLafferty & Preston, 1992).

Space plays a central role in women's employment and economic status for several reasons. The journey to work connects women's domestic lives with their paid employment, fixing in space women's multiple roles. In commuting to work, women incur direct time and monetary costs, costs that can be significant for women trying to juggle the competing demands of domestic work, child care, and paid employment, and also for the growing number of women who are sole breadwinners for their families. Second, labor market processes are articulated in space, so space can be both a barrier to, and a facilitator of, women's employment. The location of job opportunities influences the kinds of jobs available to women and women's ability to find paid employment. Because of their domestic responsibilities and exclusion from male-dominated networks, women typically search for work more locally than do men and rely more on family- and community-based contacts for employment information (Hanson & Pratt, 1991). These spatial constraints on job search make job accessibility a significant factor in women's employment and earnings. For minority women whose job search is further restricted by racial segregation and discrimination, spatial constraints can be critically important.

The first section of this chapter summarizes general trends in metropolitan commuting and describes the important differences in commuting behavior among gender and race groups. We then review the handful of studies that have examined geographic aspects of minority women's employment to show that the commuting behavior of minority women reflects their economically disadvantaged position in the labor market, their domestic roles, their access to transportation, and the pervasive impact of racial segregation. The final section emphasizes the need to take into account differences in women's experiences, especially for minority women. Their employment decisions are made within a different social, historical, and spatial context than those made by other women, and these contextual differences demand the attention of feminist scholars and geographers alike.

■ Metropolitan Commuting

Recent trends in North American travel patterns indicate that although the share of work trips relative to total travel has declined, the journey to and from work (commuting) still accounts for one fifth of all person trips annually. There are more commuting trips in the United States and Canada than ever before. In the United States, a growing share of trips are to suburban workplaces, with suburb-to-suburb flows accounting for the largest proportion (one third) of metropolitan work trips. Indeed, suburban areas now employ 60% of metropolitan workers. Similar trends are occurring in Canada, where suburb-to-suburb flows are increasing along with a general increase in the extent of metropolitan labor markets (Tranplan Associates, 1990). Residential and employment suburbanization have led to an increase in automobile travel. In addition, workers' access to private vehicles has increased in recent decades, whereas alternatives to the automobile have declined. Thus 93% of all work trips in 1980 were made by automobile. The average length of the work trip in the United States was approximately 10 miles or 22 minutes one way in the 1980s and by 1990 was 10.9 miles (Pisarski, 1987; U.S. Department of Transportation, 1991). Work trips are approximately the same length in Canada, where the average commuting time was 24 minutes in 1986 (Frederick, 1990).

Some of the variables commonly considered relevant in the literature on differences in workers' commuting lengths are mode of transportation, income, occupation, residential location, and workplace location. Travel time varies strongly by mode use. Workers who rely on public transit spend longer times getting to work than do those who use private vehicles, although auto commuters travel much longer distances than do bus or rail riders or pedestrians (Guiliano, 1979; Singell & Lillydahl, 1986). Low-income workers are less able to afford long commutes and thus work closer to home (Madden, 1981). Workers in high-status occupations also tend to commute farther than do those in clerical, sales, and service occupations (Cubukgil & Miller, 1982). Commuting differences between occupation groups are related to differences in income as well as differences in residential and employment location (Westcott, 1979). Also, the spatial distribution of workplaces within a metropolitan area leads to differences in work trips (Blumen & Kellerman, 1990; Fagnani, 1983). In particular, studies on commuting have shown links between occupational segmentation and the geographical distribution of jobs (Johnston-Anumonwo, 1988; Hanson & Pratt, 1988; Hwang & Fitzpatrick, 1992).

Gender

These factors aside, one prevalent attribute in work trip differences has been the worker's gender. The journey-to-work literature is filled with evidence that, for the population at large, women work closer to home. In 1983, the average work trip distance for women in the United States was 8.3 miles compared to 11.2 miles for men (Pisarski, 1987, p. 27). In Canada, the average time spent commuting to work was 27.5 minutes for men and 22 minutes for women (Frederick, 1990). Numerous studies have shown that women have shorter, more localized commutes than do men, in terms of both the home-work separation distance and time spent commuting to work (e.g., Gordon, Kumar, & Richardson, 1989a; Hanson & Johnston, 1985; Howe & O'Connor, 1982; Madden, 1981; Rutherford & Werkele, 1988). White (1986) also shows that the factors that affect women's commuting differ from those of men. The factors outlined earlier are responsible for women's shorter commutes: women's lower use of private vehicles, their lower wages, their low-status occupations, the spatial distribution of women's jobs, and, perhaps the most common explanation, their greater household and child care responsibilities.

Despite general agreement on women's shorter work trips, many studies indicate that working women are not homogeneous in their travel behavior (e.g., England, 1993; Hanson & Pratt, 1991). Until recently, geographers studying gender differences in urban travel behavior simply assumed that working women of all racial and/or ethnic groups have shorter work trips than do men. This assumption is unfounded in view of the large body of research that shows significant racial differences in travel patterns, as summarized later. Unfortunately, these studies have also been flawed by not paying sufficient attention to gender differences in the journey to work.

Race

The literature on racial differences in commuting and spatial access to employment is largely concerned with the spatial mismatch hypothesis. First proposed in 1968, the hypothesis states that African American inner-city residents have poorer spatial access to jobs than do other workers, because of their concentration in segregated residential areas distant from and poorly connected to major centers of employment growth. Lack of access leads to high rates of unemployment and, for persons able to overcome spatial barriers and find work, to long journeys to work. Holzer (1991) has provided a comprehensive review of the literature on the spatial mismatch hypothesis; therefore, we just briefly summarize major findings.

Representative national-level data for U.S. metropolitan areas show differences between European Americans and African Americans in their journeys to work. African Americans spend a longer time commuting to work and are less likely than European Americans to commute by car. The weight of the evidence supports the spatial mismatch hypothesis by documenting either longer commute times for African American workers or reduced access of minority workers to suburban employment centers (Cooke & Shumway, 1991; Holzer, 1991; Ihlanfeldt & Sjoquist, 1989; Leonard, 1987).

Some studies have questioned the spatial mismatch hypothesis, however. Using national data for 1977 and 1983, Gordon, Kumar, and Richardson (1989b) find no evidence that non-Whites suffer more commuting constraints. Examining only workers who commute by car, they found that non-Whites tend to travel *shorter distances* but have *longer travel times* than do Whites, irrespective of central-city or suburban residence. The authors interpret these findings as insufficient evidence of locational constraints for Blacks. Their conclusion that African American workers do not commute longer thus differs from the conclusions of many researchers who have analyzed commuting data. Ellwood's (1986) study of African American teenage employment in Chicago also questions the validity of the spatial mismatch hypothesis. He found little impact of spatial access to jobs on Black unemployment, leading him to conclude that race, not space, lies at the heart of African American unemployment.

Despite the diversity of findings on the spatial mismatch hypothesis, it is clear that African Americans differ from European Americans in their commuting patterns and spatial and social access to employment. These differences affect African American women as well as men, although women are often neglected in the spatial mismatch debate. Analyses of accessibility to employment must take into account the unique economic, social, and locational profile of African American women, a profile that both influences and is influenced by women's commuting and spatial access to work. The next section addresses how African American women's economic status, domestic roles, and locational and transportation characteristics affect their abilities to commute to and search for fulfilling employment opportunities.

■ Economic Status and Commuting

Women's commuting and spatial access to work are critically linked to their economic status. The low-wage, part-time jobs that women often hold offer little incentive for commuting long distances, and women's willingness and ability to commute affects the types of job opportunities available to them. Many studies confirm the importance of economic factors in women's shorter work trips (Madden, 1981; White, 1986). As Madden (1981) writes, "If women had the same job tenure, work hours and, most importantly, wages as their male counterparts in the household, their work trips would no longer be shorter, in fact they would be longer" (p. 191). Research indicates that for both men and women, the length of work trip increases with earnings: Higher earnings both enable and encourage long commutes (Ihlanfeldt, 1992). However, according to Rutherford and Wekerle (1988), the financial payoff from each additional mile of commuting is less for women than for men because of the shortage of high-wage job opportunities for women. Thus women's shorter work trips are in part an economically rational response to the lack of well-paid, full-time job opportunities.

Women's overrepresentation in a narrow set of "female" occupations also influences their work trip patterns. Women are heavily concentrated in the secondary sphere of the labor market, in clerical, sales, and service occupations, and this occupational segmentation is closely tied to their commuting decisions. In general, workers who hold secondary jobs have shorter work trips, mainly because of their lower incomes, so women's concentration in secondary occupations is associated with shorter commutes (Cooke & Shumway, 1991; Gordon et al., 1989a; Johnston-Anumonwo, 1988). But there are also strong links between occupational segmentation and commuting independent of wage effects. Studies of firm location decisions show that firms sometimes intentionally choose sites that are accessible to a "captive" female labor force—a labor force consisting primarily of married, White, suburban women (Nelson, 1986). On the other hand, England (1993) found little evidence of clerical firms relocating to gain proximity to a spatially entrapped female labor force. Clearly firms' motivations differ, but in some instances, proximity to a gender-defined labor force is important. Occupational segmentation is also partly an outcome of spatial factors. Jobs in "female-dominated" occupations are unevenly distributed over space, and women who desire to work near home may choose those occupations because of their closeness to home (Blumen & Kellerman, 1990; Hanson & Pratt, 1988). Hanson and Pratt (1991) found that women employed in female occupations worked closer to home, placed more value on the proximity of work to home, and searched for work more locally than other women. Thus labor market segmentation is both a cause and a consequence of women's spatial access to employment.

Researchers have only recently begun to explore the applicability of these concepts for African American women. Layered on these generalities are the powerful effects of racial discrimination in job and housing markets that uniquely circumscribe African American women's economic, social, and spatial positions. Do our generalizations about women's commuting and economic status apply in this context? The effects of wages and incomes on commuting are more complex for African American women than for European American women. Black women earn less, on average, than do White women, and, theoretically, this should lead Black women to have shorter work trips. Yet studies indicate that African American women have significantly longer work trips than do White women, especially as measured by commuting time (Johnston-Anumonwo, 1991; Madden & Chen Chui, 1990). For example, in the New York metropolitan area, McLafferty and Preston (1991) found that African American women's average commuting time was more than 10 minutes longer than European American women's, and the racial difference remained even after controlling for income, occupation, and industry of employment.

The seemingly paradoxical combination of low wages and long work trips does not mean, however, that economic factors are irrelevant in African American women's commuting and spatial access to employment. For employed African Americans, there is a strong positive relationship between wage or income levels and commuting time (McLafferty & Preston 1991, 1992; Zax, 1990). As with White workers, the length of the work trip increases with income, but for African Americans, this occurs within the confines of a strongly segregated housing stock. In Detroit, Zax (1990) found that high-income Black workers lived in predominantly Black residential areas distant from work, whereas lower-income Black workers lived in similarly segregated areas closer to work. African American's commutes to a suburban workplace were systematically longer than those of White workers, indicating a spatial mismatch between jobs and residences for African Americans irrespective of income.

Research on the relationship between occupational segmentation and commuting for minority women is in its infancy, but initial findings show important differences between White and African American women. To the extent that commuting time is indicative of spatial factors, McLafferty and Preston (1992) found that for White women, spatial factors directly affect occupational segmentation, but not for African American women. African American women who worked in occupations typical of their gender/race group did not have shorter commute times than those who worked in other occupations, indicating that proximity to home was not a factor in Black women's concentration in gender- and race-segregated occupations. These findings suggest that social and historical forces are much more important than spatial factors in explaining the occupational segmentation of African American women.

Spatial access to employment also affects African American women's ability to find work and thus their labor force participation. For African American women living in central cities, the long commutes required to access growing suburban employment centers can pose a formidable barrier to finding and keeping a job, especially for single parents

(Blackley, 1990). Research on the links between unemployment and spatial access to jobs yields mixed results about the importance of spatial effects. Some studies, most notably Ellwood's (1986) research in Chicago, show no relationship between African American unemployment rates and the local availability of jobs. On the other hand, carefully designed, disaggregate studies find that after controlling for individual and family characteristics, the spatial accessibility of job opportunities has strong effects on the likelihood of employment for African Americans (Ihlanfeldt & Sjoquist, 1990). In the latter research, spatial effects were significant for African American women as well as for men, and overall, African Americans had much poorer spatial access to employment than did Whites. Among teenagers, African American women were more likely to be unemployed than African American men with similar educational and family characteristics, a sign that young women's domestic responsibilities limit their ability to find work. Further research is needed to analyze these relationships in a variety of geographic settings and to explore in detail how spatial access affects African American women's job search processes and employment decisions.

■ Domestic Roles

Original explanations for women's shorter work trips focused on the gender division of labor within the household, emphasizing that women worked close to home to minimize commuting time, thereby making more time for child care and other domestic responsibilities (Gordon et al., 1989a; Madden, 1981). Whether the decision to work near home is a preference or a constraint is still debated (Hanson & Johnston, 1985). However, studies in the United States, Canada, Sweden, and France have all demonstrated that family circumstances have a major impact on women's travel (Fagnani, 1983; Hanson & Hanson, 1980; Rutherford & Wekerle, 1988). Single and married mothers have different travel patterns than do men and childless women (Rosenbloom, 1993).

Research has yielded mixed results about the influence of children and marital status on commuting times. Although some studies show that mothers have shorter work trips than do childless women (Fagnani, 1983; Singell & Lillydahl, 1986), others find no relationship between commuting time or distance and the presence of children (Gordon et al., 1989a; Hanson & Johnston, 1985; White, 1986). The effects of marriage are equally controversial, with only some studies demonstrating that married women travel shorter distances than do single women (Fagnani, 1983; Hanson & Johnston, 1985; Johnston-Anumonwo, 1992b).

The contradictory empirical evidence may well be due to inadequate conceptualization and measurement of domestic responsibilities. Domestic roles are examined in a fragmentary fashion, with each aspect of household composition considered separately. For example, most studies consider separately the effects of child care responsibilities, as described by the ages and presence of children and the effects of marital status. The relationships between domestic roles and women's economic status are also examined infrequently, despite preliminary evidence that the two are linked. Recent studies show that the number of wage earners in the household influences women's work trips. Women from two-worker households are more sensitive to commuting distance than are men, unlike women in single-worker households (Fagnani, 1993; Johnston-Anumonwo, 1992a; Madden, 1981). Finally, the varied and changing nature of the strategies that women use to accommodate multiple and competing domestic and employment responsibilities is often overlooked (England, 1993; Fagnani, 1993).

A comprehensive conceptualization of domestic roles and their relation to other aspects of women's lives is crucial for understanding African American women's access to employment. Apart from the fact that low wages often make long work trips unaffordable for women, obtaining a job near the residence continues to be identified as important for women who need to attend to family responsibilities (Hanson & Pratt, 1991). Yet the limited research about African American women's work trips does not support this otherwise reasonable behavior. In studies of Baltimore, Buffalo, and Rochester, Johnston-Anumonwo (1991, 1992b) finds that even when they use an automobile for the work trip, African American mothers still travel longer than do European American mothers. Among childless women, there are no racial disparities in trip length. Preston, McLafferty, and Hamilton (1993) reported similar findings in the New York metropolitan region where married African American women commuted longer than did their European American counterparts. However, in New York, there were marked racial differences in trip length among childless women.

Current research allows us only to speculate about the reasons that the presence of children has less impact on African American women's work trips. Preston et al. (1993) found that domestic roles as described by marital status and the presence of children had little impact on the commuting times of African American women. Single, childless African

American women commuted 3 minutes less on average than did their married counterparts with children at home. Among European American women, the disparity was much larger, 9.2 minutes. These relationships indicate that African American women do not adjust their work trips to accommodate domestic roles in the same ways as do European American women. The authors only speculate on the reasons for these disparities, noting that African American women may have developed strategies for accommodating domestic and employment responsibilities that do not reduce their commuting times. African American mothers rely more on relatives and less on paid child care than do European American mothers (Liebowitz, Waite, & Witsberger, 1988). The flexibility of noninstitutional day care may enable African American mothers to travel long times to work. Alternatively, the employment decisions of African American women may be so restricted by occupational segregation and discrimination that they are unable to reduce work trips when faced with competing domestic and employment demands.

■ Transportation Access

Disparities in access to transportation have a fundamental influence on the work trips of African American women. At a time when most people travel to work by car, members of households with no private automobiles suffer a relative disadvantage regarding employment access. There is a marked reduction in travel time when the automobile is the mode of travel (McLafferty & Preston, 1992; Singell & Lillydahl, 1986). In fact, people without jobs are disproportionately members of carless households (Kasarda, 1989). Among these carless households are many poor, female-headed households that cannot afford car ownership and maintenance.

The disproportionate number of African American women who are primary providers of their households' incomes and their low average wages mean that employed African American women are far more likely to rely on public transportation than are European American women. Many studies confirm African American women's greater dependence on public transit. Using data for Baltimore, Johnston-Anumonwo (1991) found that over 25% of African American women and men used public transit compared with 14% and 6% of European American women and men. In northern New Jersey, analysis of 1980 census data showed that 25% of African American women used mass transit compared with 14%

of African American men (McLafferty & Preston, 1992). Similar patterns were observed in Buffalo and Rochester, New York, where African American women were much more reliant on mass transit than either African American men or White men or women (Johnston-Anumonwo, 1992b). African American women's dependence on mass transit is a major factor in their long commuting times (McLafferty & Preston, 1992).

Just as with many other aspects of their lives, women's access to transportation is contingent on marital status, age, and family structure, as well as on income. Among households with two wage earners, men are more likely to drive to work than women (Hanson & Hanson, 1980; Singell & Lillydahl, 1986). As a result, married working women are more reliant on public transit for the work trip than are men (Fagnani, 1983; Guiliano, 1979; Wekerle & Rutherford, 1989). As yet, we have little information to assess how household structure and family composition affect African American women's access to transportation. Neither the direct effects of family composition on access to cars nor their indirect effects related to income and residential location have been examined for African American women. Given their distinctive household circumstances, particularly the preponderance of low-income, female-headed families, it is unwise and certainly inappropriate to generalize on the basis of information about the entire population.

The local context alters these relationships, particularly for African American women who live near the centers of large metropolitan areas in the United States. At the core of urban areas served by rapid transit, gender differences in commuting are small. Men and women both use public transportation to travel quickly to jobs in the central business district (Crowe, 1992; Fagnani, 1983; Preston & McLafferty, 1994). Despite their concentration adjacent to central areas, African American women do not seem to obtain the full benefit of rapid transit systems. Their commuting times generally exceed those of European American women with similar incomes and occupations (McLafferty & Preston, 1992). Residential segregation may mean that African American women live farther from job opportunities or that they rely more on slow surface transportation that increases travel times (Leonard, 1987; Rutherford & Wekerle, 1988). Recent transit development in American cities has often favored wealthy White suburban neighborhoods rather than minority populations in the inner city.

In suburban parts of North American cities, women are far more likely to drive to work than their counterparts living in central cities.

Indeed, in suburban areas, the vast majority of workers, of both sexes, drive to work (Crowe, 1992; Preston & McLafferty, 1994). Suburban reliance on cars cuts across all racial groups. In the suburbs of New York, equal proportions of African American and European American women, 75%, drive to work (Preston et al., 1993). It is noteworthy that despite the overwhelming reliance on cars by men and women living in suburban areas, gender differences in work trip times are significant and consistent (Hanson & Pratt, 1991; Preston & McLafferty, 1994). Women living in the suburbs drive shorter times to work than do their male counterparts.

The complicated pattern of gender differences in access to transportation and its effects on African American women's work trips underscores the importance of examining the local context in which men and women link their work and domestic lives. Gender differences in commuting have emerged in suburban areas where men and women are equally likely to drive to work, suggesting that relative location affects far more than access to transportation modes.

■ Location Effects

The relative locations of home and work affect access to employment in complex ways that we are only beginning to explore. However, our analyses are limited by a shortage of detailed geographical information describing the locations of home and work that can be related to individual information about labor market participation, race, gender, access to transportation, household composition, and other factors thought to influence the journey to work. As a result, the majority of studies examine access to employment at an aggregate level, comparing central and suburban areas.

In the United States, central-city locations are distinguished from non-central-city locations and suburban areas. Finer distinctions, for example, between inner and outer suburbs, are drawn occasionally, but intraurban locations are defined crudely for the most part. The primitive designation of central and suburban areas reflects the reliance on the Public Use Microdata Sample (PUMS) of the U.S. census in much of the recent research that has analyzed the influences of gender, race, and location on access to employment (Cooke & Shumway, 1991; Johnston-Anumonowo, 1992b; Madden & Chen Chiu, 1990; McLafferty & Preston, 1991). The PUMS data provide detailed individual information about labor force participation, household composition, the journey to work, and transit use coded by residential and workplace location. However, to maintain confidentiality, locational information is limited to county and subcounty areas.

There are similar problems in Canada, where data equivalent to the PUMS do not specify intraurban locations for home or workplace. As a result, municipal boundaries are often used to separate central and suburban areas (Preston & McLafferty, 1994; Wekerle & Rutherford, 1989). Rose and Villeneuve (1988) have employed a more sophisticated typology in which the inner city was defined as the area within 30 minutes commuting time of downtown; however, these detailed categories are the exception rather than the rule.

Even with these crude definitions of central and suburban areas, there is consistent evidence of a gender difference in employment locations—women are more likely than men to have localized work trips. They work more often than men do in the area of residence (Fagnani, 1983; Howe & O'Connor, 1982; McLafferty & Preston, 1992). Women are also more likely than men to work in suburban locations, especially if they also live in the suburbs (Crowe, 1992; Fagnani, 1983; Hwang & Fitzpatrick, 1992; Johnston-Anumonwo, 1988; Singell & Lillydahl, 1986). However, the extent to which women's employment is localized in the place of residence is under debate. Analyzing clerical workers in Columbus, Ohio, England (1993) reported that many women were willing to commute long distances to jobs that have relocated from central areas to the suburbs. Women gave up convenience to retain employment in firms that had moved away from their homes and neighborhoods.

One explanation for these contradictory findings may lie in the social cleavages within central residential areas. Two main groups of working women live in central locations. In gentrified neighborhoods, upper-middle-class professional women from dual-income households enjoy proximity to work and essential services, particularly day care (Fagnani, 1993; Rose, 1993). Nearby, large and growing numbers of single female heads of households are sole breadwinners for their households (Holcomb, 1984). The two groups live in adjacent but separate central neighborhoods, segregated on the basis of class, family structure, race, and ethnicity. Residential segregation within central areas means that African American women live in less desirable, peripheral neighborhoods distant from job opportunities, especially from growing suburban employment nodes (Blackley, 1990; Ihlanfeldt & Sjoquist, 1989).

Although the suburbs are no longer monolithic, if they ever were, suburban residents in American and Canadian cities are still more likely than central residents to be wealthy and living in intact nuclear families with young children (Bourne, 1989; White, 1987). Within the suburbs, residential segregation persists (Darden, 1990), creating separate geographies of employment opportunities for African American and European American women. Preliminary evidence indicates that suburban residence is beneficial for African American women's commutes. In New York, African American women who lived in suburban areas had shorter commuting times than did their counterparts living in central areas (Preston et al., 1993).

The growth of employment in the suburbs also affects African American women's access to jobs. Increasing numbers of African American workers are commuting outward to suburban workplaces (Pucher & Williams, 1992). African American workers living in the suburbs are closer to suburban jobs than are central residents who must reverse commute (Schneider & Phelan, 1990). Because women are more likely than men to be employed in suburban workplaces, the benefits of suburban residence may be greater for African American women than for African American men.

In summary, studies consistently show that African American women commute longer than do European American women, and the gender gap in commuting, so often noted in the literature, does not exist for African Americans. This review has examined research that explicitly focuses on gender and race groups and probes the causes of differences in commuting and employment access. Although most studies have been partial in their examination of a single aspect of African American women's employment in a single spatio-temporal context, certain general conclusions emerge. The most important is that African American women's long commutes are not simply a result of their economic status or domestic roles. In fact, according to conventional wisdom, African American women's disadvantaged position in the labor force and concentration in female-headed households should lead to *shorter* work trips, not longer ones. As employed mothers, African American women do not enjoy the relative convenience of short commutes to work, nor do they gain any financial payoff for enduring long work trips. Thus the causes of African American women's long work trips must lie elsewhere—for example, in their lack of access to private transportation or their poorer social and spatial access to jobs. More research is needed to untangle the effects of these factors in different spatial and temporal contexts.

■ Agenda for Future Research

Understanding the spatial dimensions of African American women's employment requires theoretical and empirical research on a broad range of topics. By examining the complex interactions between African American women's economic and social status, access to transportation, and spatial access to employment, we will gain a more complete description of the barriers to employment and determinants of commuting decisions. Existing studies have focused narrowly on a single aspect of African American women's employment, while ignoring other key dimensions. For example, we still do not know why after controlling for differences in economic and household characteristics, African American women commute longer than do European American women. Multivariate and multidimensional studies, which analyze many factors simultaneously, will allow us to sort out the importance of transportation and spatial effects and consider their relationships to women's economic and social status.

Most existing research has also been limited in its spatial and temporal coverage. Researchers have typically examined older industrial cities in the United States and Canada, and it is not clear if these findings are applicable to most metropolitan areas. Other cities should be studied, especially those in the South and West, which have different economic bases and markedly different spatial structures from the deindustrialized cities of the Northeast and Midwest. In terms of time frame, the reliance on 1980 census data makes many studies out of date. As noted earlier, the 1980s were an important decade for African American women, a decade that saw a decline in Black women's earnings relative to White women's and continued economic restructuring and suburbanization of employment. How have these changes affected African American women's ability to seek paid employment and shoulder rising domestic and economic responsibilities? Analyzing the impacts of these changes on African American women's employment is a critically important agenda for future research.

Such multifaceted research in diverse metropolitan settings will also provide a more informed basis for urban policy making. African American women's earnings and employment are critically important to the well-being of families in metropolitan areas, yet, often, women's employment needs are ignored in the design of urban policies. We know that the lack of transportation options and shortage of available job opportunities are significant barriers to African American women's employment. Carefully designed housing, transportation, and employment information policies are needed to overcome these barriers and give African American women the same employment opportunities as European American men and women. In addition, although research indicates that working African American women have devised creative strategies to cope with child care and domestic responsibilities, clearly, providing affordable, conveniently located child care services is essential for improving women's incomes and employment options.

Our findings emphasize the need to take into account differences in women's experiences, especially for minority women. The broad social cleavages defined by race, ethnicity, and class cut across gender divisions, complicating our understanding of women's employment. Feminist theories must come to grips with this diversity if they are to be fully relevant to women's lives. Although we have stressed differences between women based on race, it is equally important to avoid racial stereotypes by addressing diversity *within* race groups. African American women lead as varied lives as do European American women, and it is important to look within social groups, not just between them. Describing and examining this diversity requires detailed, disaggregate studies of African American women's employment and commuting decisions. How do African American women search for work, and how does geographical access to job opportunities affect their employment decisions? How do African American women cope with child care and other domestic responsibilities while participating in the paid labor force? How have economic restructuring and cutbacks in social spending affected African American women's access to work? Addressing these questions will provide a richer understanding of the social and spatial dimensions of minority women's employment and the strategies women adopt in integrating home and work in time and space.

REFERENCES

Amott, T., & Matthaei, J. (1991). *Race, gender and work: A multicultural economic history of women in the United States*. Boston: South End Press.

Blackley, P. (1990). Spatial mismatch in urban labor markets: Evidence from large U.S. metropolitan areas. *Social Science Quarterly, 71*, 39-52.

Blumen, O., & Kellerman, A. (1990). Gender differences in commuting distance, residence and employment location. *Professional Geographer, 42*, 54-71.

Bourne, L. (1989). Are new urban forms emerging? Empirical tests for Canadian urban areas. *Canadian Geographer, 33*, 312-328.

Brewer, R. (1988). Black women in poverty: Some comments on female-headed households. *Signs: Journal of Women in Culture and Society, 13,* 331-339.

Cooke, T., & Shumway, J. (1991). Developing the spatial mismatch hypothesis: Problems of accessibility to employment for low-wage central city labor. *Urban Geography, 12,* 310-323.

Crowe, J. (1992). *Gender differences in commuting to office concentrations within the Toronto area.* Master's thesis, York University, North York, Ontario.

Cubukgil, A., & Miller, E. (1982). Occupational status and journey-to-work. *Transportation, 11,* 252-276.

Darden, J. (1990). Differential access to housing in the suburbs. *Journal of Black Studies, 21,* 15-22.

Ellwood, D. (1986). The spatial mismatch hypothesis: Are there teenage jobs missing in the ghetto? In R. B. Freeman & H. Holzer (Eds.), *The Black youth employment crisis* (pp. 147-187). Chicago: University of Chicago Press.

England, K. (1993). Suburban pink collar ghettos: The spatial entrapment of women? *Annals of the Association of American Geographers, 83,* 225-242.

Fagnani, J. (1983). Women's commuting patterns in the Paris region. *Tidjschrift voor Economische en Sociale Geografie, 74,* 12-24.

Fagnani, J. (1993). Life course and space, dual careers and residential mobility among upper middle class families in the Ile-de-France region. In C. Katz & J. Monk (Eds.), *Full circles: Geographies of women over the life course* (pp. 171-187). New York: Routledge.

Frederick, J. (1990, Winter). Commuting time. *Canadian Social Trends,* pp. 29-30.

Gordon, P., Kumar, A. & Richardson, H. (1989a). Gender differences in metropolitan travel behavior. *Regional Studies, 23,* 499-510.

Gordon, P., Kumar, A., & Richardson, H. (1989b). The spatial mismatch hypothesis: Some new evidence. *Urban Studies, 26,* 315-326.

Guiliano, G. (1979). Public transportation and the travel needs of women. *Traffic Quarterly, 33,* 607-616.

Hanson, S., & Hanson, P. (1980). Gender and urban activity patterns in Uppsala, Sweden, *Geographical Review, 70,* 291-299.

Hanson, S., & Johnston, I. (1985). Gender differences in work-trip lengths: Explanations and implications. *Urban Geography, 6,* 193-219.

Hanson, S., & Pratt, G. (1988). Spatial dimensions of the gender division of labor in a local labor market. *Urban Geography, 9,* 180-202.

Hanson, S., & Pratt, G. (1991). Job search and the occupational segregation of women. *Annals of the Association of American Geographers, 81,* 229-253.

Holcomb, B. (1984). Women in the rebuilt urban environment: The U.S. experience. *Built Environment, 10,* 18-24.

Holzer, H. (1991). The spatial mismatch hypothesis: What has the evidence shown? *Urban Studies, 28,* 105-122.

Howe, A., & O'Connor, K. (1982). Travel to work and labor force participation of men and women in an Australian metropolitan area. *Professional Geographer, 34,* 50-64.

Hwang, S., & Fitzpatrick, K. (1992). The effects of occupational sex segregation and the spatial distribution of jobs on commuting patterns. *Social Science Quarterly, 73,* 550-564.

Ihlanfeldt, K. (1992). Intraurban wage gradients: Evidence by race, gender, occupational class and sector. *Journal of Urban Economics, 32,* 70-91.

Ihlanfeldt, K., & Sjoquist, D. (1989). The impact of job decentralization on the economic welfare of central city Blacks. *Journal of Urban Economics, 26*, 110-130.

Ihlanfeldt, K., & Sjoquist, D. (1990). Job accessibility and racial differences in youth employment rates. *American Economic Review, 80*, 267-276.

Johnston-Anumonwo, I. (1988). The journey to work and occupational segregation. *Urban Geography, 9*, 138-154.

Johnston-Anumonwo, I. (1991, October). *Sex differences in the racial gap in commuting.* Paper presented at the meeting of the Middle States Division of the Association of American Geographers, State College, PA.

Johnston-Anumonwo, I. (1992a). The influence of household type on gender differences in work trip distance. *Professional Geographer, 44*, 161-169.

Johnston-Anumonwo, I. (1992b, October). *Racial and gender differences in work trip patterns.* Paper presented at the meeting of the Middle States Division of the Association of American Geographers, Syracuse, NY.

Kasarda, J. (1989). Urban industrial transition and the urban underclass. *Annals of the American Academy of Political and Social Science, 501*, 26-47.

Leonard, J. (1987). The interaction of residential segregation and employment discrimination. *Journal of Urban Economics, 21*, 323-346.

Liebowitz, A., Waite, L., & Witsberger, C. (1988). Child care for preschoolers: Differences by child's age. *Demography, 25*, 205-220.

Madden, J. (1981). Why women work closer to home. *Urban Studies, 18*, 181-194.

Madden, J., & Chen Chui, L. (1990). The wage effects of residential location and commuting constraints on employed married women. *Urban Studies, 27*, 353-369.

McLafferty, S., & Preston, V. (1991). Gender, race and commuting among service sector workers. *Professional Geographer, 43*, 1-15.

McLafferty, S., & Preston, V. (1992). Spatial mismatch and labor market segmentation for African-American and Latina women. *Economic Geography, 68*, 406-431.

Nelson, K. (1986). Female labor force supply characteristics and the suburbanization of low-wage office work. In A. Scott & M. Storper (Eds.), *Production, work and territory* (pp. 149-171). London: Allen and Unwin.

Pisarski, A. (1987). *Commuting in America: A national report on commuting patterns and trends.* Westport, CT: Eno Foundation.

Preston, V., & McLafferty, S. (1994). Gender and employment in service industries: A comparison of two cities. In F. Frisken (Ed.), *The changing Canadian metropolis* (pp. 123-149). Berkeley: University of California Press.

Preston, V., McLafferty, S., & Hamilton, E. (1993). The impact of family status on Black, White and Hispanic women's commuting. *Urban Geography, 14*, 228-250.

Pucher, J., & Williams, F. (1992). Socioeconomic characteristics of urban travelers: Evidence from the 1990-91 NPTS. *Transportation Quarterly, 46*, 561-581.

Rose, D. (1993). Local childcare strategies in Montreal, Quebec: The mediations of state policies, class and ethnicity in the life courses of families with young children. In C. Katz & J. Monk (Eds.), *Full circles: Geographies of women over the life course* (pp. 188-207). New York: Routledge.

Rose, D., & Villeneuve, P. (1988). Gender and the separation of employment from home in metropolitan Montreal, 1971-81. *Urban Geography, 9*, 155-179.

Rosenbloom, S. (1993). Women's travel patterns at various stages of their lives. In C. Katz & J. Monk (Eds.), *Full circles: Geographies of women over the life course* (pp. 208-242). New York: Routledge.

Rutherford, B., & Wekerle, G. (1988). Captive rider, captive labor: Spatial constraints on women's employment. *Urban Geography, 9,* 116-137.

Schneider, M., & Phelan, T. (1990). Blacks and jobs: Never the twain shall meet? *Urban Affairs Quarterly, 26,* 299-312.

Singell, L., & Lillydahl, J. (1986). An empirical analysis of the commute to work patterns of males and females in two-earner households. *Urban Studies, 23,* 119-129.

Statistics Canada. (1990). *Women in Canada: A statistical report (2nd ed.).* Ottawa, Ontario: Minister of Supply and Services.

Tranplan Associates. (1990). *Trip diary survey analysis.* Ontario: Ministry of Transportation.

U.S. Department of Transportation. (1991). *1990 nationwide personal transportation study, early results.* Washington, DC: Federal Highway Administration.

Wekerle, G., & Rutherford, B. (1989). Employed women in the suburbs: Transportation disadvantage in a car-centered environment. *Alternatives, 14,* 49-54.

Westcott, D. (1979). Employment and commuting patterns: A residential analysis. *Monthly Labor Review, 102,* 3-9.

White, M. J. (1986). Sex differences in urban commuting patterns. *American Economic Review, 76,* 368-373.

White, M. J. (1987). *American neighborhoods and residential differentiation.* New York: Russell Sage.

Zax, J. (1990). Race and commutes. *Journal of Urban Economics, 28,* 336-348.

14

Public Housing and the Beguinage

DAPHNE SPAIN

One of the ways in which gender has been placed on the urban research agenda in the planning profession is through the transformation of public housing from its origins as a way station for married couples into housing of last resort for unmarried women. In the process, public housing has become a "dreadful enclosure" (Walter, 1977) of poverty resistant to federal and local efforts to improve living conditions for residents and reduce welfare dependency. Because over three quarters of public housing residents now are women, public housing has become a distinctly gendered urban problem.

In searching for solutions to contemporary problems, it is often useful to turn to historical precedent. The purpose of this chapter is to illustrate similarities (and occasional differences) between American public housing in the late 20th century and the medieval *beguinage*. The beguinage was a quasi-secular urban institution for unmarried women too poor to join convents; it arose in response to the skewed sex ratio created by men's participation in the Crusades. Both public housing and the beguinage are urban gendered spaces (Spain, 1992) that hold "surplus" women in abeyance until socially sanctioned roles are created for them (Mizruchi, 1983).

Creating spaces predominantly for women was not the original intent of public housing, as it was for the beguinage. Yet both have warehoused women who live outside the married-couple ideal (and, in the case of public housing, increasingly outside a changing labor market). By housing unmarried poor women in such places, society contains spatially some of

AUTHOR'S NOTE: In addition to Judith Garber and Robyne Turner, thanks are due Bob Beauregard, Kim Hopper, Peter Marcuse, Hal Mizruchi, Steven Nock, and Lisa Reilly for comments on an earlier draft of this chapter.

the problems associated with the growth of female householders: poverty, out-of-wedlock pregnancies, and a general threat to the patriarchal social order.

■ Controlling Surplus Populations

In *Regulating Society*, sociologist Ephraim Mizruchi (1983) argues that societies create "abeyance structures" to regulate the impact of surplus populations by holding potentially dissident, marginal people under surveillance in controlled situations until "status vacancies" open elsewhere. He cites monasteries, beguinages, Bohemian communes, compulsory apprenticeship/education, and the Works Progress Administration (WPA) writers's and artists's projects as examples of institutions that evolved in response to the pressures of large numbers of marginal and potentially disruptive people at different points in history.

Formal schooling, for example, emerged in 19th-century England after apprenticeships had declined and simultaneously with an increase in the number of young persons. The ambiguous status of young adults created the threat of uncontrolled behavior and required some form of warehousing until they were needed in the labor market. In America, WPA projects employing writers and artists were a small part of New Deal legislation aimed at "regulating the poor" until the economy improved after the Great Depression (Mizruchi, 1983, p. 112; Piven & Cloward, 1971). Abeyance structures thus mitigate the impact of dissident behavior and slow the rate of social change.

Unmarried women are often marginal and sometimes dissident. Because the proper roles for women historically have been those of wife and (married) mother, women falling outside those boundaries have presented problems for the social order. Women in cities, especially, have been identified with disorder and danger since Juvenal attributed the decline of Rome to female immorality (E. Wilson, 1991, p. 15).

Beguines were neither fish nor fowl, occupying the status of "secular nun"; they were pious but uncloistered women, some of whom had children and most of whom were employed in the local economy (Neel, 1989). Beguinages attracted and held thousands of women who might otherwise have jeopardized a social order built on the patriarchal ideal as espoused by the church (Mizruchi, 1983, pp. 62-65).

Although it is more socially acceptable for women to be unmarried in contemporary America than it was in medieval Europe, female householders still present a threat to American family values as espoused by conservatives such as Charles Murray (1984). The possibility that poor unmarried women could also become dissidents was demonstrated when welfare mothers created the (then) radical National Welfare Rights Organization in the 1960s (Piven & Cloward, 1971) and when low-income women working as community activists in the 1980s challenged the structure of federal welfare programs (Naples, 1991).

■ Contemporary Demographics

In a different era, the "status vacancies" to which unmarried women might eventually aspire would be that of wives. But the realities of household formation in the late 20th century make it unlikely that poor female householders (especially Blacks) will find or choose husbands.

Several factors have contributed to the rise in the numbers and proportions of unmarried women. The proportion of women in their early 30s who never marry rose from 7% to 16% between 1960 and 1990. Divorce rates have risen, as has the incidence of out-of-wedlock births. The proportion of White women with premarital births rose from 9% to 22%, compared with a rise from 42% to 70% for Black women between 1960 and 1989. Women now head over 1 in 10 White families and nearly 1 out of 2 Black families (U.S. Bureau of the Census, 1992).

Women maintaining families alone are more likely to be living in poverty (32%) than are married-couple families (6%) (U.S. Bureau of the Census, 1991, p. 6). Higher poverty rates mean that female-headed families are more likely to be eligible for subsidized housing. In 1989, three quarters of public housing households were headed by women, compared with 42% of all renter households nationally; Blacks constituted 53% of public housing households compared with 18% of all renter households (Casey, 1992, pp. 21-22). Striking racial differences also exist *within* public housing: Housing for the elderly is overwhelmingly White, whereas Blacks predominate in family projects (Bickford & Massey, 1991).

Explanations for the higher incidence of Black than White female householders include the skewed sex ratio for African Americans. Some analysts identify the shortage of African American men of marriageable age (due to high rates of incarceration and homicide) relative to the number of eligible women as the reason for the large proportion of female householders among Blacks (Bennett, Bloom, & Craig, 1989;

Espenshade, 1985; Tucker, 1987). Others emphasize the characteristics of men, rather than their numbers, as potential deterrents to married-couple household formation. High rates of unemployment and low educational attainment among Black men, for example, place constraints on marital opportunities for Black women (Spanier & Glick, 1980; Tucker & Taylor, 1989). Still others believe that the combination of a skewed sex ratio and poor economic opportunities for Black men has contributed to the growth in Black female householders (South & Lloyd, 1992; W. J. Wilson, 1987).

A similarly skewed sex ratio existed during the Middle Ages for Europeans. Warehousing solutions then were provided, in part, by the beguinage.

■ The Medieval Beguinage

The religious fervor of the 12th and 13th centuries took its toll on the European population. The combined effects of the Crusades and the growth of monasteries resulted in an unbalanced sex ratio, especially in urban areas. Because women had few economic options outside of marriage, the shortage of men presented significant problems. Daughters from wealthy families could join a religious order, but daughters of peasants or artisans could not afford the dowry required to enter a nunnery (Power, 1975, p. 89). Thus the beguinage emerged as a temporary (and for some, a permanent) domicile for women who had left their families of origin but had not yet married. It gave otherwise marginal women an acceptable identity during a period of transition.

Women known as Beguines occupied roles somewhat between the sacred and secular. The Franciscan Gilbert of Tournai remarked in 1274 that "there are among us women whom we have no idea what to call, ordinary women or nuns, because they live neither in the world nor out of it" (Neel, 1989, p. 323). Their marginal status gave Beguines independence, otherwise impossible in medieval culture, from male authority in marriage and in the church.

Whether they lived singly or in groups, Beguines' households excluded men. Most Beguines lived in urban communities informally connected with one another (Neel, 1989). Beguinages were formed when wealthy individuals donated city houses for the use of women workers. The Great Beguinage of Namur, Belgium, for example, was founded in 1235 when the widow Eve donated three houses near St.

Aubain for use by the Beguines. Namur eventually had five operating beguinages (McDonnell, 1969, p. 69).

By the 13th century, beguinages had been loosely organized around the model of nunneries, although the Beguines operated independently of the church and were never cloistered. Some historians, in fact, identify beguinages more closely with women's guilds than with religious orders because their residents pursued a variety of crafts (such as spinning, weaving cloth, and "gold beating") in addition to caring for the sick and needy (Neel, 1989, p. 324). By the 14th and 15th centuries, ecclesiastical authorities created autonomous parishes of beguinages, granting them civil and religious status. One of the secular privileges extended to beguinages was exemption from local taxes (Boulding, 1976, pp. 445-447; McDonnell, 1969, pp. 5, 6, 85).

By the 14th century, beguinages had assumed the character of poorhouses where unmarried women and their children could find shelter, food, and sometimes work. Beguinages formed at these later dates were called "domus pauperum" (McDonnell, 1969, pp. 82, 88). Whether one sees the beguinage as workers' guild or poorhouse, its connection with religious activity was never entirely absent. Beguines, like other parishioners, were required to obey their priests. They were more independent than nuns (writing their own psalters and installing their own female priests and confessors) but could not afford to flaunt the authority of the church too rashly because the price of heresy was burning at the stake (Boulding, 1976, p. 449).

Constant efforts were made to bring the Beguines more closely in line with church doctrine and with acceptable standards of conduct for women. In 1319, Robert, Count of Flanders, sought to regulate the status of Beguines by recommending that only women "of condition" be allowed to live in beguinages. "Of condition" meant respectable and sound moral character, as well as possession of some material wealth (McDonnell, 1969, p. 96).

The Beguine movement produced several leaders of historical note. Mary d'Oignies organized a group of men who then created support within ecclesiastical structures for the *Frauenbewegung* (women's movement) that existed outside religious orders. Christine Stemmeln was another woman whose male disciples supported women's activities. Mechtild of Magdeburg and Beatrice of Nazareth lived all their lives in beguinages and contributed greatly to the religious culture of the day. More important than these few remarkable women, however, were the thousands who earned livings and established a status for themselves in society by virtue of their association with the beguinage (Boulding, 1976, pp. 449-453).

Before turning to parallels with public housing, it is important to point out one significant difference between the beguinage and public housing as gendered abeyance structures: Beguines were theoretically celibate, whereas a large proportion of unmarried women in public housing have children. Although the presence of children presents different demands on the type of shelter provided and on the ability of all residents to work, I would argue that the actual presence of children is less relevant to the comparisons proposed here than their representation as reminders of uncontrolled sexuality. Because Beguines were not cloistered, their sexual behavior could not be monitored; indeed, there is evidence that beguinages contained children in the 14th century (McDonnell, 1969, pp. 82, 88). Likewise in the late 20th century, the abolition of punitive eligibility and eviction criteria has resulted in the inability of housing authorities to monitor the sexual behavior of public housing tenants. Thus the relative (although not absolute) absence of children in beguinages and their presence in public housing should not obscure useful comparisons between the two institutions.

■ Twentieth-Century Public Housing

Peter Marcuse (1993) has pointed out that the term *public housing* can be historically misleading because it represents differences in the timing of and reasons for construction, by whom and for whom it was built, its design, and its site location. Focusing only on the buildings themselves allows us to define public housing as those units built, bought, rehabilitated, and/or managed by housing authorities established by the Wagner-Steagall Act (also known as the U.S. Housing Act) of 1937. Construction proceeded sporadically from the 1930s until the 1980s, when budget cuts effectively eliminated new production (Fisher, 1959; Mitchell, 1985). In 1989, there were 1.3 million units of public housing (4% of the total rental stock), two thirds of which were located in central cities and three quarters of which were occupied by unmarried women (Casey, 1992).

The first eligibility criteria for public housing clearly favored married couples. Even before the Wagner-Steagall Act was passed, the George-Healey Act of 1936 specified that public housing should be available only to *families* who lacked sufficient income to afford decent, safe housing in the private market. Occupancy was further restricted to "responsible tenants, chosen in order that 'every family accepted for consideration is capable of paying the rent and has a satisfactory character' " (Fisher, 1959, p. 88). Because three quarters of all households consisted of married couples in 1940 (Bianchi & Spain, 1986, p. 88) and "responsible tenants with satisfactory character" were most likely to be married couples by the definitions of local officials, the George-Healey Act both responded to prevailing demographics and reinforced norms regarding the unsuitability of unmarried households for public housing.

When the 1937 Housing Act was created, the emphasis on families continued. The type of family was clearly defined in the U.S. Housing Authority's 1938 pamphlet titled "What the Housing Act Can Do for *Your* City":

> The immediate purpose of public housing is to raise the living standards of typical *employed* families of very low income, who are independent and self-supporting. . . . Public housing is designed to improve the condition of millions of working families who have reasonably steady jobs and reasonably steady but inadequate earnings. (Fisher, 1959, p. 164, emphasis in original)

Not until 1956 were public housing benefits extended to elderly single persons and to families defined as consisting of two or more persons ordinarily related by blood, marriage, or adoption—a tacit recognition of unmarried mothers and their children (Fisher, 1959, p. 94).

Early proponents of public housing were aware of its potential analogy with the poorhouse and vociferously decried its association with relief recipients. Edith Elmer Wood, author of five highly regarded books on housing published between 1914 and 1935, worked diligently with Catherine Bauer and others lobbying the federal government to build well-designed, low-income housing supportive of family life (Birch, 1978). In an influential 1939 article titled "One Third of a Nation," Wood announced " 'in no uncertain terms' " that public housing should avoid fostering " 'colonies of dependents' " (Bauman, 1987, p. 53).

The National Association of Housing Officials (NAHO) was equally clear in its 1939 instructions to local housing authority administrations. Its *Manual From Early Experience* contained a section on "Relations With Local Government Agencies." Listed under "Special Relationships With Welfare Agencies" was the advice that welfare agencies and the local housing authorities should limit the number or percentage of relief families. NAHO believed that the "concentration of public welfare families in a housing project would immediately place a brand upon the development and would prevent the establishment of a normal community representing various income levels and normal occupations" (NAHO, 1939, pp. V21-V22).

Given that one of the goals of public housing was to encourage upward mobility, surprisingly little data exist on residential turnover in public housing units. One exception is the national "average monthly move-out rates" for 1944 to 1957 computed by Fisher (1959, p. 169). The 1944 rate of 1.29 move-outs per month per 100 public housing units had increased to 2.19 monthly move-outs by 1957. Whites experienced greater mobility than did Blacks. Between 1950 and 1959, national mobility data were recorded separately for all-White and all-Black projects. White move-out rates exceeded Black move-out rates in every year by more than 2 to 1 (Fisher, 1959, p. 169). Racial discrimination in the private market, fewer resources with which to compete, and demolition of their neighborhoods during urban renewal meant that poor Blacks in public housing had few places to move *to* during the 1950s.

Contemporary data suggest that public housing tenancy has become more permanent. Public housing residents had the longest tenure of any renters in 1989: Half had lived in their current units for 5 years or more compared with only one quarter of all renters. One quarter of all public housing tenants had lived in the same unit for 10 or more years compared with 13% of all renters (Casey, 1992, p. 21). Although it is not possible to compare data for the 1950s directly with that for the 1980s, it appears that public housing was more effective as a transitional housing solution (especially for Whites) early in its history than it is now.

"Problem families" began to emerge in the early 1950s. Among the problem characteristics listed by the Philadelphia Housing Authority in 1952, for example, were included people who " 'engage in prostitution, or random relationships which resemble it.' " Because men are seldom accused of prostitution, the Housing Authority was undoubtedly referring to the behavior of unmarried (and possibly married) women. The Housing Authority also considered asking prospective tenants to present a valid marriage license to pass a "social criteria" test—that is, of a "normal" husband-wife couple. Admitting the wrong kind of family could destroy the mission of public housing, because " 'a chronic alcoholic,

TABLE 14.1   Characteristics of Households Living in Public Housing, for
                 Selected Years (in percentages)

Household Characteristics	1952	1957	1989
Female headed	25	27	76
Black	40	48	53
Receiving welfare	29	38	45
Earning wages/salaries	74	69	35

SOURCES: Data for 1952 and 1957 comes from Fisher (1959, p. 168). Data for 1989 comes from Casey (1992, pp. 21-24).

wife-beater, or a *woman with several children and generally unaccept-*
able behavior [italics added] will create such tensions among many
surrounding families that the objective of [public] housing will be
defeated in many ways' " (Bauman, 1987, p. 134). Like the Count of
Flanders's effort to restrict beguinages only to women "of condition,"
early public housing officials wanted to restrict their units to women of
sound moral character.

The ability of the Philadelphia Housing Authority to control "problem families" ended in 1968 when they were sued by Oliva Byrd, an
unwed mother who had been denied housing. The Pennsylvania Supreme Court ruled that the Housing Authority could not "exclude or
indifferently treat" any family solely because it included children born
out of wedlock. This ruling reduced the power of the Housing Authority
to regulate the sexual behavior of tenants (Bauman, 1987, p. 203).

In addition to less control over the private lives of tenants, housing
authorities were undergoing financial pressures in the 1960s created by
inadequate federal funding. Tenants who were previously paying fairly
competitive rents, some at significant proportions of their income, were
protected by the Brooke Amendments of 1969 and 1971 from paying
more than 25% of their incomes for rent (eventually raised to 30%
beginning in 1982). This legislation drastically reduced the operating
budgets of housing authorities and increased their financial burdens.
Deprived of the ability to generate revenues from higher rents, housing
authorities began to accept welfare payments as a means of guaranteeing
income (Mitchell, 1985, p. 196). Table 14.1 demonstrates the changing
profile of public housing tenants between 1952 and 1989.

Changes in eligibility criteria ushered in an era of the government's
assuming a greater support role for female-headed families. The number
of families receiving Aid to Families With Dependent Children (AFDC)

benefits nearly tripled between 1960 and 1970 and continued to expand throughout the 1980s. By 1983, 1.3 million (predominantly female-headed) households were receiving both a housing and income subsidy. Approximately 22% of the welfare population received a direct housing subsidy, and approximately 38% of households receiving housing subsidies also received income assistance (Newman & Schnare, 1988, pp. 50-51; U.S. Department of Health and Human Services, 1990, p. 37). By 1989, nearly one half of public housing tenants were receiving welfare (see Table 14.1).

Some of the early welfare policies seemed deliberately designed to reinforce women's reliance on (and surveillance by) the government. In the 1950s, various states enacted "suitable home," "substitute father," and "man in the house" rules that disqualified large numbers of Black and unwed mothers. By 1960, one half of all states had a "suitable home" policy. They used these rules to deny aid to single mothers who took in men as boarders, cohabited with men, or refused to identify the fathers of children born out of wedlock. "Substitute father" rules identified any man in the mother's life as responsible for her children's support. Special investigative teams conducted "midnight raids" to search for a man who, if found, automatically became the substitute parent. Discovery of a man in the house could result in the loss of benefits for both women and children (Abramovitz, 1988, pp. 323-324).

The history of welfare is a history of regulating women's lives in ways that have ensured that single mothers will remain poor and dependent on the government (Abramovitz, 1988). The effects of public housing eligibility criteria, combined with national changes in household composition trends and fiscal policies affecting housing authorities, transformed public housing from a place in which married couples predominated to a gendered space consisting of poor women. This shift has been accompanied by a decline in the proportion of public housing tenants earning wages and an increase in the proportion receiving welfare.

■ Conclusion

Several similarities exist between the beguinage and contemporary public housing. Beguines were dependent on the church in much the same way that public housing tenants are dependent on the government. Ecclesiastical authorities arranged local tax exemptions for the beguinage;

the federal government exempts public housing from local taxes. The church provided poor relief for Beguines; the government provides AFDC to nearly one half of all public housing tenants. Relations between the Beguines and the church were always ambivalent in terms of the degree of control that could be exerted over the beguinage. The history of public housing eligibility criteria reflects a comparable tension between contemporary government officials and public housing residents. In short, the relationship between each abeyance structure and the dominant institution of the era is similar.

Several differences exist between the beguinage and public housing. Most obvious is the greater presence of children in public housing. I believe, however, that children are the visible reminders of the government's inability to control the sexuality of female householders in public housing, just as the church was unable to completely control the sexuality of Beguines during the Middle Ages. Other differences include the manner in which women enter and exit the institution and the degree of stigma attached to each type of shelter.

Beguinages and public housing also have differed in their ability to "demarginalize" unmarried women. Beguines worked to support themselves and thus were directly engaged with the local economy. Public housing has not required work of its residents, but recent efforts to sell public housing units to tenants (Connerly, 1986; Rohe & Stegman, 1992) could be interpreted as an attempt to create recognized status for single women as homeowners. Given that such programs have proven most effective for two-earner households and in scattered site locations, however, it is unlikely that they would be effective for the poorest female householders in concentrated developments. If public housing is an abeyance structure from which poor women are unlikely to exit via marriage or home ownership, a third alternative for demarginalization is engagement in the labor force. The women who have become most successful in public housing have done so by creating jobs for themselves and for others; their stories have taken on a mythical aura, much as the stories of Beguines who successfully reached beyond the boundaries of the beguinage.

Kimi Gray, for example, was a welfare mother of five living in Washington DC's Kenilworth Parkside public housing development when she began organizing tenants for management roles. The Kenilworth Parkside Resident Management Corporation eventually became a multimillion-dollar enterprise that created jobs for residents and reduced school dropout rates, teen pregnancies, and crime (U.S. House of Representatives, 1985). Bertha Gilkey created a similar success in St. Louis. The Cochran Tenant Management Corporation supports jobs and housing construction for residents of the Cochran Gardens housing development (DeParle, 1992). By 1986, 14 other public housing developments had formed similar management organizations (Weisman, 1992, pp. 112-113).

The common threads uniting the beguinage and successful public housing developments are engagement with the labor market and strong internal leadership. Beguines wove cloth and spun tapestries; tenants in Kenilworth Parkside and Cochran Gardens run beauty shops, health centers, moving companies, and child care facilities. Kimi Gray and Bertha Gilkey may be the modern equivalents of Mechtild of Magdeburg and Beatrice of Nazareth.

It is important, of course, to reexamine the type of productive activity that counts as work. Mundane domestic tasks such as doing the laundry can be effectively converted to an entrepreneurial enterprise that provides a service and job training for public housing residents simultaneously, as it has in Nickerson Gardens in Los Angeles (Leavitt, 1993). Similarly, the presence of children in public housing provides the opportunity for day care employment.

Small businesses are not the only solution to the demand for employment among public housing tenants, but they can arise organically in response to community needs. They may also help fill the gap between short-term and long-term efforts to reduce welfare dependency. In Baltimore, for example, the Family Development Center (FDC) in Lafayette Courts coordinated and delivered a range of on-site services for public housing residents. Although participation in the FDC program did not increase employment rates, it did increase participation in the Job Partnership Training Act's preparation for employment (Shlay & Holupka, 1992). Given the isolation of many public housing developments from existing jobs and transportation (Bauman, Hummon, & Muller, 1991; Vobejda, 1993), locally created businesses may be the most promising avenue to employment for female householders in public housing.

This chapter began with the observation that public housing, like the beguinage, is an institution holding women of marginal marital status in abeyance until alternative roles open to them. Whether female householders are as much a threat to the social order as medieval Beguines is debatable. The greater likelihood that public housing residents will have children out of wedlock may make them more dangerous to the patriarchal ideal than Beguines, for whom celibacy was the norm. Whether similarities between the institutions outweigh the differences between them is also a matter of debate. I propose that their shared function of providing shelter for unmarried women and their relationship to the dominant institutions of the day make the beguinage and public housing more similar than different.

All the problems associated with contemporary public housing cannot be solved by reference to earlier models for housing poor women in cities. Yet the beguinage, a place demographically similar to public housing, appears to have relevant lessons for the present. Connecting unmarried women directly to the economy through employment—rather than through a husband's job or through government subsidies—may hold as much promise for the incorporation of "marginal" women into acceptable statuses now as it did during the Middle Ages. History has provided the precedent. It is up to planners to use that history to define what is possible for the future.

REFERENCES

Abramovitz, M. (1988). *Regulating the lives of women: Social welfare policy from colonial times to the present.* Boston: South End Press.

Bauman, J. F. (1987). *Public housing, race, and renewal: Urban planning in Philadelphia, 1920-1974.* Philadelphia: Temple University Press.

Bauman, J. F., Hummon, N., & Muller, E. (1991). Public housing, isolation, and the urban underclass: Philadelphia's Richard Allen Homes 1941-1965. *Journal of Urban History, 17,* 264-292.

Bennett, N. C., Bloom, D. E., & Craig, P. N. (1989). The divergence of Black and White marriage patterns. *American Journal of Sociology, 95,* 692-722.

Bianchi, S., & Spain, D. (1986). *American women in transition.* New York: Russell Sage.

Bickford, A., & Massey, D. (1991). Segregation in the second ghetto: Racial and ethnic segregation in American public housing, 1977. *Social Forces, 69,* 1011-1036.

Birch, E. L. (1978). Woman-made America: The case of early public housing policy. *Journal of the American Institute of Planners, 44,* 130-144.

Boulding, E. (1976). *The underside of history: A view of women through time.* Boulder, CO: Westview.

Casey, C. (1992). *Characteristics of HUD-assisted renters and their units in 1989.* Washington, DC: U.S. Department of Housing and Urban Development, Office of Policy Development and Research.

Connerly, C. E. (1986). What should be done with the public housing program? *Journal of the American Planning Association, 53,* 142-155.

DeParle, J. (1992, January 5). Cultivating their own gardens. *New York Times Magazine,* pp. 22-26, 32, 46, 48.

Espenshade, T. (1985). Marriage trends in America: Estimates, indicators, and underlying causes. *Population and Development Review, 11,* 193-245.

Fisher, R. M. (1959). *Twenty years of public housing.* New York: Harper & Brothers.

Leavitt, J. (1993). Shifting loads: From hand power to machine power to economic power. *Intersight, 2,* 53-57.

Marcuse, P. (1993, October). *Interpreting "public housing" history.* Paper presented at the annual meeting of the Association of Collegiate Schools of Planning, Philadelphia.

McDonnell, E. (1969). *The Beguines and beghards in medieval culture.* New York: Octagon.

Mitchell, J. P. (1985). Historical overview of direct federal housing assistance. In J. P. Mitchell (Ed.), *Federal housing policy and programs* (pp. 187-206). New Brunswick, NJ: Rutgers University Center for Urban Policy Research.

Mizruchi, E. (1983). *Regulating society: Marginality and social control in historical perspective.* New York: Free Press.

Murray, C. (1984). *Losing ground: American social policy 1950-1980.* New York: Basic Books.

Naples, N. (1991). Contradictions in the gender subtext of the War on Poverty: The community work and resistance of women from low income communities. *Social Problems, 38,* 316-332.

National Association of Housing Officials. (1939). *Local housing authority administration: A manual from early experience* (Publication No. N107). Chicago: Author.

Neel, C. (1989). The origins of the Beguines. *Signs: Journal of Women in Culture and Society, 14,* 321-341.

Newman, S., & Schnare, A. (1988). *Subsidizing shelter: The relationship between welfare and housing assistance.* Washington, DC: Urban Institute Press.

Piven, F. F., & Cloward, R. (1971). *Regulating the poor: The functions of public welfare.* New York: Random House.

Power, E. (1975). *Medieval women.* Cambridge, UK: Cambridge University Press.

Rohe, W., & Stegman, M. (1992). Public housing homeownership: Will it work and for whom? *Journal of the American Planning Association, 58,* 144-157.

Shlay, A., & Holupka, C. S. (1992). Steps toward independence: Evaluating an integrated service program for public housing residents. *Evaluation Review, 16,* 508-533.

South, S., & Lloyd, K. (1992). Marriage opportunities and family formation: Further implications of imbalanced sex ratios. *Journal of Marriage and the Family, 54,* 440-451.

Spain, D. (1992). *Gendered spaces.* Chapel Hill: University of North Carolina Press.

Spanier, G., & Glick, P. (1980). Mate selection differentials between Whites and Blacks in the United States. *Social forces, 58,* 707-725.

Tucker, M. (1987). The Black male shortage in Los Angeles. *Sociology and Social Research, 71,* 221-227.

Tucker, M., & Taylor, R. J. (1989). Demographic correlates of relationship status among Black Americans. *Journal of Marriage and the Family, 51,* 655-665.

U.S. Bureau of the Census. (1991). Poverty in the United States: 1988 and 1989. *Current Population Reports* (Series P-60, No. 171). Washington, DC: Government Printing Office.

U.S. Bureau of the Census. (1992). Households, families, and children: A 30-year perspective. *Current Population Reports* (Series P-23, No. 181). Washington, DC: Government Printing Office.

U.S. Department of Health and Human Services. (1990). *Fast facts and figures about Social Security.* Washington, DC: Government Printing Office.

U.S. House of Representatives. (1985). *Barriers to self-sufficiency for single female heads of families* (Hearings before a subcommittee of the Committee on Government Operations). Washington, DC: Government Printing Office.

Vobejda, B. (1993, February 12). At core of the problem lies absolute isolation. *Washington Post*, p. A1

Walter, E. V. (1977). Dreadful enclosures: Detoxifying an urban myth. *European Journal of Sociology, 18,* 151-159.

Weisman, L. K. (1992). *Discrimination by design: A feminist critique of the manmade environment.* Urbana: University of Illinois Press.

Wilson, E. (1991). *The sphinx in the city: Urban life, the control of disorder, and women.* Berkeley: University of California Press.

Wilson, W. J. (1987). *The truly disadvantaged.* Chicago: University of Chicago Press.

15

Concern for Gender in Central-City Development Policy

ROBYNE S. TURNER

Cities define downtown success by their profitability and often follow strategies that bring middle- and upper-class persons back to downtown, defining the livability of downtown. The needs and living conditions of existing inner-city dwellers are not a primary concern of city government when making downtown development policy and selecting specialty uses for downtown land. Yet women who live in and around the downtown need the same services and outlets that suburban women need—grocery and drug stores, clothing and children's stores, and other services. In particular, downtown women who are poor need to have access to transportation, day care, and safety. Answering their needs contributes little to the economic viability and return on investment of downtown development.

Low-income women are not part of the decision-making process in most downtowns, yet some cities' development policies provide an array of living conditions and options that are of benefit to those women. The question this chapter raises is how downtown development policy advances or diminishes the availability of a range of issues that are important to low-income women residents of downtown areas, especially when those policies are developed in a political climate of urban growth.

This chapter concentrates on why cities are likely to offer an array of living options that give women flexibility in addressing their individual living goals and concerns, and how specific aspects of the

AUTHOR'S NOTE: I wish to thank Margaret Wilder and Dorothy Stetson for their helpful comments on drafts of this chapter, and Trip Cioci for assistance in data collection.

271

development policy contribute to that outcome. Spatial development patterns affect residential location choices, and the mechanisms for participation affect the scope of the development policy agenda. Thus the decision-making criteria used in downtown development will affect the breadth of downtown living options available to low-income women. Urban political economy implies that development decisions affect social groups based on class and race (Fainstein, 1987; Wilson, 1987). Gender, then, should also be used to uncover the consequences of spatial development patterns and land use options for women marginalized by the performance of the urban political economy—a significant urban group.

Policies that emphasize the development of the quality of life, such as the protection of residential neighborhoods, access to mass transportation, public participation, and flexible zoning, can provide a more hospitable environment for economically vulnerable, inner-city, female-headed households. In contrast, policies that emphasize commercial development as a market-based solution to increase downtown development are likely to give priority to the economic viability of an urban area and diminish the likelihood that low-profit alternative living options will be available.

This chapter explores the needs and characteristics of low-income women who live in and around the core of four Sunbelt central cities and the downtown development policies and decision processes that most significantly affect them. The ability of conventional development policy approaches to address these women's concerns is considered as well as the prospect of success of alternative economic development policies. Each city's policy approach and politics are examined in terms of ability to deliver alternative living arrangements that would benefit low-income women. The final part of the chapter considers why cities vary, including whether women elected to office can make an impact on the nature of downtown development outcomes and whether alternative policy processes such as neighborhood planning can make a difference.

■ The "Free-Market" Development of Urban Places

The economic restructuring of cities during the post-Fordist era significantly affected the spatial array of land uses in downtown, resulting in a developmental approach that uses land value as the basis of policy selection. This produces a gendered downtown, because it does not explicitly consider the impact of development issues on women. Cities that provide development policies based on alternative criteria are more likely to offer an array of living options and devote more attention to the quality of living conditions. This result will more likely reflect the desires of all women, but especially low-income women who will benefit from mass transit, high-density convenience and access to services and support networks, and a variety of housing options and costs. In actuality, such a variety represents an ungendered development agenda—many different living arrangements catering to different life-styles, incomes, cultures, and household types.

It is unclear whether women should present a distinct agenda to policymakers or whether they can benefit from concerns raised by other groups, such as those that concentrate on issues of neighborhood preservation, social services, and employment training. It is likely, however, that the agenda of issues raised by low-income women is more likely to reflect a comprehensive approach to living conditions and arrangements than those specific issues that smaller groups will raise.

The living conditions of low-income, inner-city women are particularly affected by patterns of downtown development, opportunities to participate in local decision making and agenda building, and the criteria used to guide development and the quality of life. Women may not have the political power to maximize their living options across these factors and, instead, may rely on leveraging other political power as a means to affect development policy. For instance, women will benefit from policies that maintain neighborhoods and affordable housing. However, unless those policies are motivated to sustain a level of living conditions, not just as a by-product of development patterns, those living conditions may evaporate as demands for capital investment opportunities increase downtown. Therefore, it is important that a variety of criteria are used to evaluate downtown development so that low-income women realize a full array of living arrangements.

Residential Concerns of Low-Income, Inner-City Women

Women of limited economic means need development strategies that provide a low-cost, high-service environment (Cook, 1988; Ritzdorf, 1986). Women are more likely than men to be concerned about living close to jobs and services, such as day care and health care, or having the transportation to get to jobs and services (Shlay & DiGregorio, 1985). Women want to have access to friends and family, particularly as a support system for single-parent households (Cook, 1988). Black women may be more satisfied with neighborhood conditions than White women if a social network and family are nearby (Thomas & Holmes, 1992). High-density housing such as that offered in the inner-city affords that network. Given the disproportionate level of Black, female-headed, poverty-level households in downtown areas, minority women are likely to need policies that offer these inner-city living conditions.

Transportation needs often force women to opt for inner-city locations where public transit is more readily available than it is in the suburbs and where services are more likely to be concentrated (Salem, 1986). The remaining poor women residents of a gentrified neighborhood may no longer be able to afford that same neighborhood and may be forced to relocate. Those women as heads of households are less likely to own a home and more likely to have lower-quality housing (although lower-cost housing) than men, especially because women are involuntary movers (Spain, 1990). Given women's lower wages (Jones & Kodras, 1990), they are likely to spend a disproportionate level of their income on housing costs (see S. Smith, 1990).

Zoning laws may restrict the availability of day care, low-cost housing, and convenient employment that low-income women need (Ritzdorf, 1986). Rapidly changing development patterns may have the same affect. Redevelopment changes the character of the downtown area, concentrating on the wants of daytime workers, resulting in a closed downtown at night and on weekends. Displaced low-income women are faced with accepting shrinking zones of low-quality downtown area housing as commercial revitalization claims the core area. Wilson (1987) suggests that this concentration actually spatially and socially isolates inner-city residents from mainstream market opportunities, leaving them vulnerable to extended unemployment. For minority women who are a single head of household, this isolation leaves few opportunities to improve their status of living (Rolinson, 1992). Although the inner-city may be less costly and more accessible than the suburbs for low-income women, it is fraught with difficulty in terms of time spent on transit and quality of housing. It is also fraught with danger in terms of the exposure to crime and violence for both women and their children.

A compounding factor in meeting the needs of inner-city, low-income women is the degree to which their earning potential is affected by economic restructuring (Northrop, 1990). The structural changes in the urban economy from 1970 to 1980 indicate a loss of family-wage union jobs as women are increasingly opting for single-household status. In addition, the rise in the low-wage service sector occurs as the number of female-headed households in poverty increases. Women that have significant family responsibilities may be less likely to take on long commutes to work and are forced to accept jobs near their inner-city residence (Hansen & Pratt, 1991).

Sunbelt Downtown Land Use

As Sunbelt inner cities and downtowns struggle to build commercial value, it is unlikely that their development policies will give significant attention to the living conditions of female-headed households (see Turner, 1993). Women suffered as economic participants during the economic restructuring of the Sunbelt (Jones & Kodras, 1990). The Sunbelt had the highest increase of all U.S. regions in service sector growth from 1970 to 1980, which significantly contributed to the increase in the feminization of poverty in that region. Black women were particularly affected, perhaps due in part to spatial isolation in the downtown area. The limited political voice of low-income women in these inner cities is outstripped by the enormous pressure to make opportunities for capital investment through land use development.

It is reasonable to expect that women heading poverty-level, single-parent households would not benefit from the growth-dominated development agendas in Sunbelt cities. There is a concentration of women as heads of disadvantaged households in the downtown area in these four Sunbelt cities (see Table 15.1). Demographic characteristics from the 1990 U.S. census comprise a set of gender distress indicators (GDIs) that portrays the proportionate presence of distressed female families in a defined location. The indicators are families as a percentage of households, female-headed families as a percentage of all families, and female-headed families below poverty with the head between the ages of 15 and 65. In addition, the overall percentages of minority populations are presented as context indicators. Data were gathered for the census tracts making up the urban central business district and the immediately surrounding area to assess the presence of these households in the downtowns of four cities: Dallas, Houston, San Diego, and Orlando. The citywide GDIs are similar across these cities. In addition, the downtown conditions in each city reflect more distress than at their respective citywide level.

The data in Table 15.1 reveal that the conditions of female distress vary across the downtown areas. Downtown development and land use

TABLE 15.1 1990 Gender Distress Indicators for Women in Downtown Areas (in percentages)

Indicators	Orlando—12 census tracts		San Diego—11 census tracts		Dallas—7 census tracts		Houston—16 census tracts	
	Census Tract	Citywide	Census Tract	Citywide	Census Tract	Citywide	Census Tract	Citywide
Female-headed families (age 15-65) below poverty	41.4	30.8	46.4	26.2	62.8	29.8	52.4	32.2
Families headed by female	32.5	24.3	25.8	17.7	52.6	22.8	37.4	22.7
Households that are families	42.5	56.5	38.0	62.0	44.6	59.2	55.0	63.4
Context indicators								
Black	32.6	26.9	10.8	9.3	57.6	29.4	37.6	28.0
Hispanic	7.1	8.7	51.6	20.6	9.6	20.8	45.5	27.6

SOURCE: U.S. Census, 1990. Construction by author.

policies may contribute to these differences, although the limited number of places in this study prevents statistical analysis or definitive conclusions. However, the working premise of this comparison is that low-income women are constrained and isolated in the inner-city by development policies and strategies used by the city. The ability to create a climate where ungendered development concerns are heard in the land use decision process, resulting in attention to living conditions, is key to mitigating urban gendered distress.

■ Availability of Living Options

Women-headed households increasingly make up the basis of inner-city households, and research suggests that they have identifiable life-style interests. The value of these interests, however, is not adequately dealt with by traditional downtown economic development measures. The problem lies in using profit potential as the standard criteria to evaluate commercial development as a function of the urban political economy. Instead, conditions that benefit low-income women reflect a quality of life that may not be easily evaluated by economic criteria, such as increased property value.

The availability of a variety of living options for low-income women downtown reflects a sustainable quality of urban development. Sustainable development in the inner city is not necessarily defensible based on economic contribution but does represent a standard of living in terms of physical land use, availability of services, and housing options. Yet, traditionally, the downtown quality of life is measured by the presence of professional sports arenas and cultural amenities because they enhance the attractiveness of the city as a good business climate indicator. A downtown quality of life so defined is not likely to create conditions of importance to low-income women, such as day care and mass transportation, because those uses may not necessarily enhance the value of downtown property.

Low-income women and their households likely are overlooked in development policy unless there is political intervention to disrupt the process that continues to rely on economic production values to drive land use policy making.

The living conditions of low-income, inner-city women are also limited by social controls inherent in the policies for development and land use that contribute to the spatial restructuring of urban places (Beauregard, 1989; N. Smith, 1990; Soja, 1989). These controls limit the availability and costs of residence locations, types of housing, employment opportunities, and services such as transportation, day care, and health care. There is little demand to stop commercial and residential gentrification in downtown given the significant exchange value of downtown land (Logan & Molotch, 1987).

Instead, downtown development will offer living conditions that attract professional single women and women in "empty-nest" households, who will contribute to the economic profitability of downtown by consuming increasingly valued properties and who are drawn to quality-of-life investments such as cultural amenities. This dichotomy puts women on a collision course defined by class, perpetuating the use of economic criteria as the appropriate guide to development.

Uneven Urban Development and Women's Living Options

The interests of low-income women are marginalized by the degree to which economic criteria direct downtown development in much the same way that class and race are considered according to their economic impact on downtown. Generally, the public and private sectors initiate development after determining its potential economic contribution, stimulating growth and facilitating commercial construction demands. Critiques of the focus on cost-benefit analysis of development suggest that it overlooks the potential advantages and disadvantages to minorities and poor persons who populate the downtown area (Keating & Krumholz, 1991; Squires, 1989). Although women inhabit downtown areas, their interests are not considered as explicitly in these analyses. The literature does not emphasize the value of the relationship of social issues and living conditions that add to the quality of life of inner-city women.

The downtown development process is not motivated to address the living conditions of low-income women residents, yet some cities enact policies that create alternative living options. Although the political influence of land-based interests captures the development policy process (Mollenkopf, 1983; Stone, 1989; Swanstrom, 1985) and government acts either as a well-meaning facilitator of those interests, hoping to stimulate development for the benefit of a wide range of interests (Stone & Sanders, 1987), or as a facilitator that is fiscally dependent on the success of those free-market interests (Peterson, 1981), the literature suggests that local governments can and do make political choices enabling these alternative development policies (Vogel & Swanson, 1989; Turner, 1992). Therefore, political pressure on the development policy process can ensure that a variety of living arrangements downtown are available.

Some cities have opted to use linkage policies to balance the impact of new development with the needs of downtown residents and businesses (Lassar, 1989). This approach extracts development contributions to provide living options important to low-income women, among others. Affordable housing, job training programs, day care facilities, social services, and transportation are made available (see Dreier & Ehrlich, 1991; Goetz, 1989; Herrero, 1991). Linkage policies represent a positive contribution to women's living conditions by subsidizing necessary services that low-income women cannot afford. More important is that this subsidization represents a deliberate effort by the city through its development policy to override the capitalist cost-benefit-driven outcome that would normally preclude the availability of such services. The danger for low-income women, however, is that linkage policies designed to provide a measure of equity to living downtown may disappear if the market reacts negatively to the costs of implementing those requirements (Frieden & Sagalyn, 1989).

Access to the decision process is another factor that affects whether development policy addresses the living conditions of low-income women through alternative development criteria. Development partnerships and linkage policies often operate within the confines of development authorities and quasi-governmental decision processes, often for the purpose of minimizing political opposition. Low-income women likely will not be able to affect the future of their neighborhoods under those conditions. Inner-city women have less ability to control their living conditions than do suburban women, where living conditions are a policy priority. In decision making about downtown areas, public participation is limited. Low-income women have little recourse if economic profitability determines the parameters of downtown development. The policy process perceives these women as economic dependents, not capital contributors, which limits their access to political recourse. The political process does not accommodate or recognize their claims as legitimate or as contributions to the economic performance of downtown. A gendered understanding of development policy is needed in addition to class- or race-based explanations. Gender-based insights are an alternative to the political-economy-driven understanding of urban development policy.

■ Downtown Development Approaches in Four Cities

The approaches of four Sunbelt cities to downtown development represent significant contrasts in the effects on the quality of life for low-income women residents and provide insights into the application of both the traditional economic criteria and alternative criteria to downtown development. Houston and Dallas typify the reliance on market-based criteria to guide development, whereas San Diego and Orlando reflect a greater attention to alternative criteria by incorporating neighborhood-based planning to guide development. The implications for low-income women of these development policies are compared across the four cities.

The more that low-income women and/or their concerns achieve political access and power in the development process, the better their living conditions should be. Open, participatory systems should serve to elevate their priorities. An important consideration is that although low-income women's interests may be represented, it is the pervasiveness of those interests over the long term that indicates the level of attention for gendered concerns.

Dallas and Houston

The public booster approach to development found in Dallas and Houston has several implications for the living conditions of low-income women residents of these cities. In Dallas, separate interests were not encouraged to participate in the political process. Women were not taken seriously in this forum, although other minority political groups were successful in gaining policy concessions for their votes. Only the heads of corporations were invited to belong to the Dallas Citizens Council (DCC), which acted as a power elite, slating and electing business candidates to city offices (Elkin, 1987). Women precluded from the boardroom were precluded from the center of political power in Dallas, as were their interests. The DCC nominated a woman in the late 1950s who won a seat on city council (Fairbanks, 1990). She represented the business community, not women's interests, and there was not a substantive call to change the Dallas development agenda. Those who controlled the political process did not encourage an open dialogue on the demands of various segments of the city. Access to alternative agendas for development is limited by the closed nature of the power structure. The city vigorously resisted any factionalism or overtures to interest groups, including minority voting strength through district elections, ostensibly for the economic good of the city (Fairbanks, 1990).

Dallas's approach to Deep Ellum, an inner-city historic Black neighborhood provides insight about the consequences for low-income women residents of using economic criteria to guide development. Deep Ellum suffered from economic neglect and physical disruption from new expressways constructed to bisect the area. This historic area, although bawdy and excessive in its heyday, represented the center of Black commerce and social ties in Dallas, just northeast of downtown. Recently, the city tried to revive the area as an incubator neighborhood and arts district, to complement the revived cultural, government, and financial districts downtown (Sloan, 1983). Neither living conditions nor the availability of housing was considered in the plans to revive Deep Ellum under the guise of an economic stimulus for downtown.

Dallas displaced existing in-town affordable public housing for new upscale housing for downtown professionals in the State-Thomas area (Tatum, 1985). The State-Thomas development project addresses many of the concerns of low-income women, such as access to public transportation and a concentration of personal and retail services. However, the development is not affordable. Consequently, low-income women have not directly benefited from any efforts to revitalize the downtown.

In stark contrast to downtown experiences was an intense campaign and court battle to prevent the destruction and force the improvement of public housing in West Dallas, just across the river from downtown (Suchman & Lamb, 1991). This area did not have services, grocery stores, day care, or transportation, yet its residents (primarily low-income women) actively sought to prevent its demise because the area provided a close proximity to the rest of the city, including downtown; an established social network; and affordable housing (although poor quality). It was a neighborhood that acted as an important support system to low-income, single-female heads of households. That the city of Dallas would not entertain such needs in its approach to revitalizing downtown development suggests the impact that political power and participation have on the ability to direct attention to gender concerns.

Houston's development experience and the level of attention given to the living conditions of low-income women is similar to Dallas's experience but with a more profound reliance on the economic-value criteria for making land use and development decisions downtown. Houston provided for few alternative points of view on development priorities in that there was little public access to the development decision making for downtown. Where Dallas used civic pride as a leverage to discourage alternative views, Houston developers felt no need to address critics because their process for development decision making was strictly viewed as the domain of the private landowners and businesses downtown (Parker & Feagin, 1990). This limited access made it unlikely that any quality-of-life criteria to address the living conditions of poor women would guide downtown land use and development.

The gap in understanding between the city and local residents over downtown area development is quite evident in the decisions made concerning a proposed commercial development, Founder's Park, and the neighborhoods that were subject to displacement, Allen Parkway Village and particularly Freedman's Town. City officials, including Mayor Kathy Whitmire, were willing to displace these historic Black areas to implement privately made development decisions. Low-income women dominated the area and lived in privately held rental units and public housing.

Similar to the West Dallas experience, low-income women valued these areas as neighborhoods that provided a sustainable living environment for them. A low-income woman who was a longtime resident of

Freedman's Town headed the community group to help create and sustain small businesses, such as a self-service laundry, to show the viability of the area to the residents. The dimensions of the quality of life in Freedman's Town centered on social networks, the affordability of housing, and its proximity to other necessary services and facilities, including downtown (Henson, 1991a, 1991b).

The interests particular to low-income women did not fare any better in Houston than in Dallas. Tenant and neighborhood organizations in both Allen Village and Freedman's Town used state and federal means to halt the proposed commercial redevelopment. The closed-decision process in both cities limited the political viability of gender concerns and reflected an unwillingness to accept anything other than economic value as the criteria by which land development decisions were made.

San Diego and Orlando

The interests of low-income women in San Diego and Orlando are represented by these cities' willingness to facilitate citizen participation, preserve a variety of neighborhood opportunities in-town, and publicly address options for residential locations, amenities, and employment opportunities. This represents a wider availability of living arrangements and attention to gender concerns than was evident in Dallas and Houston. The planning emphasis is more inclusive of a broader constituency, normally limited to downtown developers. The San Diego and Orlando city governments have played a strong role in addressing the desires of neighborhood interests, including low-income areas surrounding the downtowns, along with the demands of private development interests.

In Orlando, low-income women have access to the development agenda through the city's efforts to include neighborhood resident participation in growth management planning. The city has taken steps to preserve and enhance the residential aspect of its wide array of downtown neighborhoods through its growth management plan. Affordable and minority neighborhoods are included in the downtown area strategy along with historic and upscale neighborhoods. The city employs linkage policies and bonus options to encourage more affordable housing citywide as well as zoning that expressly limits commercial encroachment on residential neighborhoods downtown (Turner, 1992).

The level of inclusiveness belies a city where the development conditions are ripe for a growth machine. Yet the city exhibits evidence that economic criteria guide its neighborhood redevelopment strategies. An example is Holden, a low-income Black neighborhood on the edge of downtown. Although it benefited from the development policies of Orlando, the city emphasized the economic development of its business community through the Holden business group, led by male business owners. Day care, grocery services, and other quality-of-life issues important to low-income women have not surfaced. Although neighborhood preservation is a priority, the emphasis on economic-value criteria limits attention to gender concerns.

Low-income women may have a better chance of stimulating attention to gender concerns in the development agenda in San Diego where neighborhoods have had a strong political presence. The fortunes of several mayors, including Maureen O'Connor, were wedded to the proliferation and city attention to neighborhood groups and minorities (Richardson, 1986; Shaughnessy, 1986). The downtown development plan overtly includes several in-town neighborhoods, of varying types and incomes. The issues of interest to low-income women, however, have been addressed as residential amenities to promote the viability of downtown housing. There has been an effort to include affordably priced housing downtown and in neighborhoods near downtown (at minimum to comply with the California requirement to provide a 1:1 replacement of housing displaced by commercial and upscale revitalization).

San Diego mandates citizen participation in its land use planning process, including the Centre City Plan for downtown. Citizen groups are active political participants, most notably the Citizens Coordinate for Century 3 (Dower, 1990), who maintain the routes to access decision making and open up opportunities for alternative criteria to be considered to guide downtown development. The city recognizes the value of alternative uses in the downtown area, as evidenced by the willingness of the business-led downtown redevelopment authority, headed by a woman, to bring the Horton Plaza mixed-use marketplace development to downtown. Supporting issues such as neighborhood viability, mass transportation, and relative housing affordability are central to the planning process.

■ Why Cities Vary in Their Approaches to Downtown Development

In Dallas and Houston, the private sector dominates the planning process for downtown. Because it shuns zoning, Houston has been recognized as a leader in free-market development. It is a bastion of business dominance with an economic and land development agenda that does not regularly consider the interests of in-town, minority residents (Bullard, 1987; Feagin, 1989). Dallas, although not as free-wheeling, has relied on the steady influence of business leaders through various organizations to guide not only the selection of candidates for office but also the planning policy for downtown (Elkin, 1987). The emphasis in both of these Texas cities has been to rely on economic-value criteria to guide downtown development.

San Diego and Orlando have taken an alternative approach to planning and development, although they, too, have experienced rapid growth. San Diego has political ties to both the growth machine and the neighborhood-based slow-growth movement. The result has been land use planning that emphasizes residential uses, neighborhood identification, and the provision of a variety of housing units across income levels (Centre City Planning Committee, 1990). Orlando has followed a similar planning strategy to preserve downtown neighborhoods while responding to demands for commercial development and construction of new public facilities in the core area (Orlando Department of Planning and Development, 1988). Both of these cities politically respond to growth interests but include neighborhood interests in the process and provide for alternative land uses as well as commercial development.

San Diego and Orlando openly consider alternative criteria to guide development, giving an opening to express the concerns of low-income women. A variety of living arrangements are available in San Diego and Orlando, but attention to gender concerns is limited. However, the potential for sustained attention is greater in those places than in Dallas and Houston. The demographic data for San Diego and Orlando indicate lower GDIs in their downtowns than in Houston and Dallas. The application of the planning process and the nature of the distribution of political power may not have prevented the displacement of low-income women from downtown by commercial redevelopment. However, these cities' attention to neighborhoods and a more inclusive political arena may indicate that social networks, transportation, services, and affordable housing are available in other parts of the cities. San Diego and Orlando take policy approaches that open more location options to low-income women than the deteriorated sections of downtown that suffer from economic disinvestment. By offering neighborhood concentrations of services and housing, the city overtly attends to the quality of living conditions as its basis for neighborhood development, which also acts to provide residential locations that are alternatives to the forced convenience of downtown.

This explanation might account for why Dallas and Houston have high concentrations of distressed female households downtown. The options for poor women may be few in those Texas cities, and the downtown is, relatively, the best place for such households to find low-cost housing, concentrations of services, and transportation access. The Texas cities may not offer options for women to easily locate in other parts of the city because economic criteria continue to guide development. Given their market-oriented, closed political processes, the results are not likely to change.

The implication of such a lack of attention to gendered issues in spatial restructuring is to allow local governments to publicly support urban redevelopment for economic purposes, whereas the impact is to control the social order by controlling land use. Harvey (1989) states that the capture of "strategic spaces . . . confers a certain power over the processes of social reproduction . . . and social power relations" (pp. 186-187). The control of the land development process affects social relationships that exist in neighborhoods, residential patterns, and job development—conditions that are highly influential on the quality of low-income inner-city women residents' lives.

Harvey (1989) is perhaps most explicit by stating that the result of uneven development is a "permanent tension between the *appropriation* and use of space for individual and social purposes and the *domination* of space through private property, the state, and other forms of class and social power" (p. 177). Women as heads of poverty-level, inner-city households are not likely to be the beneficiaries of uneven development unless some intervention is applied to expressly address their concerns by assuring that an array of living options are made available in the inner city.

A final issue to explore is whether elected officials have an impact on the priorities of the development process and the resulting availability of living options for low-income women. Mayoral candidates in all four cities received support from low-income neighborhoods, yet the development variation demonstrates that political support does not result in a sustained place on the development agenda. The development process and its relationship to the political structure must be affected as well.

Women Mayors and Ungendered Agendas

Women were elected to mayoral positions in all four case cities with three of the four in office during the time period of study. This suggests political climates that may provide significant attention to the living conditions of inner-city women. Despite having women mayors, the emphasis on living conditions affecting low-income women varied across these cities, even though all the women mayors were supported by minority and neighborhood constituencies. A woman mayor does not ensure attention to gender-based concerns, especially if the local political structure is not amenable to consider alternative development strategies.

Kathy Whitmire was elected mayor of Houston in 1981 and left office after a record five terms in 1991. Annette Strauss was elected mayor of Dallas in 1987 and served through 1991. Maureen O'Connor served in San Diego from 1985 to 1993 and was succeeded by another woman, Susan Golding. Glenda Hood is currently serving as mayor of Orlando. Mayors O'Connor, Whitmire, and Strauss included issues that would affect the living conditions of inner-city women residents on their campaign agendas. O'Connor, in San Diego, ran on a neighborhood platform and gave priority to making neighborhoods livable. This agenda potentially increased the locational choices for low-income women by diversifying the types of housing and amenities available in and around the downtown area. She gave neighborhoods a priority in the development of downtown, which potentially increases the likelihood that services, amenities, and support for a variety of neighborhood users will be addressed. Neighborhoods in and near downtown serve a wide range of income levels, although the mayor's slow-growth agenda may have contributed to soaring land values and housing costs.

Mayor Whitmire in Houston also enjoyed political support from neighborhoods at all income levels. As a candidate, she ran outside of the "golden boys" network of insiders in previous administrations. However, her attention to living conditions centered on the public sector's provision of quality-of-life issues, such as art and culture, open space, and parks. That support was generally connected to enhancing the value of private property, not alternative criteria. Whitmire presided over a significant period of commercial development downtown that displaced available affordable housing, including single-room occupancy units.

Strauss in Dallas campaigned to open the decision process to a more diverse group of people (Strauss, 1992), although she was considered part of the political elite. She won her election with the help of minority voters, but the nature of decision making and the development agenda did not change appreciably. She concentrated on guiding cultural development downtown with little ability to shift political attention to issues such as living conditions.

The election of women as mayors in these three cities suggests that women have made strides in entering political arenas dominated by economic investment and development. However, the ability to fully affect that agenda of development and to add criteria that enhance living conditions is not consistent across these cities. Additional barriers to gendered concerns that enhance the attention given to locational choices, participation, and overall quality of life for low-income women exist within the politics of the development process itself.

■ Conclusion

Women have limited access to the political power that determines downtown development policies, particularly in a climate of growth politics. If women do not have access to political power, they cannot openly affect development policies, nor is it likely that they will have access to the informal routes commonly exercised by the growth machine (Molotch, 1976). The residential-location options for women and their attendant living conditions are affected by downtown development policy approaches. The gender-based literature has given little attention to gender concerns raised by development policy approaches. Although limited, this brief review of downtown development in four Sunbelt cities indicates that low-income women may affect development contours, to include alternative policy approaches and to evaluate development criteria. The expressed needs and interests of low-income women, particularly as single heads of households are most likely served by policies that enable a variety of locational choices, enhance participation to broaden the city's development agenda to include various perspectives, and offer housing and access across a broad spectrum of income and lifestyles.

The impact of policy choices on low-income women should be addressed with the same depth and clarity as are class- and race-based interests in the literature on urban political economy. Given the identification of particular interests of women in urban places, the criteria of gender should be added to the analysis. The concentration of distressed female households in inner cities warrants such inclusion in the analytical discussion of how cities develop. Whereas the growth machine and development decision arenas are dominated by male actors, increasingly, the recipients of those policy options are female. It is important to recognize the implications for women, as the heads of households, in the debate on economic restructuring, land based economies, and the portrayal of political power.

REFERENCES

Beauregard, R. (1989). Space, time, and economic restructuring. In R. Beauregard (Ed.), *Economic restructuring and political response* (Urban Affairs Annual Reviews, Vol. 34, pp. 149-179). Newbury Park, CA: Sage.

Bullard, R. (1987). *Invisible Houston*. College Station: Texas A&M University Press.

Centre City Planning Committee. (1990). *Centre City San Diego preliminary community plan*. San Diego, CA: City of San Diego.

Cook, C. (1988). Components of neighborhood satisfaction. *Environment and Behavior, 20*, 115-149.

Dower, R. (1990, April 30). C3 leader works to preserve vision of local paradise. *San Diego Business Journal*, pp. 10-11.

Dreier, P., & Ehrlich, B. (1991). Downtown development and urban reform. *Urban Affairs Quarterly, 26*, 354-375.

Elkin, S. (1987). State and market in city politics. In C. Stone & H. T. Sanders (Eds.), *The politics of urban development* (pp. 25-51). Lawrence: University Press of Kansas.

Fainstein, S. (1987). Local mobilization and economic discontent. In M. P. Smith & J. Feagin (Eds.), *The capitalist city* (pp. 323-342). Oxford, UK: Basil Blackwell.

Fairbanks, R. (1990). The good government machine. In R. Fairbanks (Ed.), *Essays on sunbelt cities and recent urban America* (pp. 125-150). College Station: Texas A&M University Press.

Feagin, J. (1989). *Free enterprise city*. New Brunswick, NJ: Rutgers University Press.

Frieden, B., & Sagalyn, L. (1989). *Downtown, Inc.* Cambridge: MIT Press.

Goetz, E. (1989). Office housing linkage in San Francisco. *Journal of the American Planning Association, 55*, 66-77.

Hansen, S., & Pratt, G. (1991). Job search and the occupational segregation of women. *Annals of the Association of American Geographers, 81*, 229-253.

Harvey, D. (1989). *The urban experience*. Baltimore, MD: Johns Hopkins University Press.

Henson, S. (1991a, May 17). Razing history. *The Texas Observer*, pp. 9-12.

Henson, S. (1991b, July 12). Allen Parkway Village. *The Texas Observer*, pp. 1, 6-14.

Herrero, T. (1991). Housing linkage. *Journal of Urban Affairs, 13*, 1-20.

Jones, J. P., & Kodras, J. (1990). Restructured regions and families: The feminization of poverty in the U.S. *Annals of the Association of American Geographers, 80*, 163-183.

Keating, D., & Krumholz, N. (1991). Downtown plans of the 1980s. *Journal of the American Planning Association, 57*, 136-152.

Lassar, T. (1989). *Carrots & sticks*. Washington, DC: Urban Land Institute.

Logan, J., & Molotch, H. (1987). *Urban fortunes*. Berkeley: University of California Press.

Mollenkopf, J. (1983). *The contested city*. Princeton, NJ: Princeton University Press.

Molotch, H. (1976). The city as growth machine. *American Journal of Sociology, 82*, 309-331.

Northrop, E. (1990). The feminization of poverty. *Journal of Economic Issues, 24,* 145-160.

Orlando Department of Planning and Development. (1988). *Downtown growth management area plan* (revised). Orlando, FL: City of Orlando.

Parker, R. E., & Feagin, J. (1990). A "better business climate" in Houston. In D. Judd & M. Parkinson (Eds.), *Leadership and urban regeneration* (pp. 216-238). Urban Affairs Annual Reviews, 37. Newbury Park, CA: Sage.

Peterson, P. (1981). *City limits.* Chicago: University of Chicago Press.

Richardson, J. (1986, February). The rise and fall of Roger Hedgecock. *California Journal,* pp. 105-108.

Ritzdorf, M. (1986). Women and the city. *Urban Resources, 3,* 23-27.

Rolinson, G. (1992). Black, single female-headed family formation in large U.S. cities. *Sociological Quarterly, 33,* 473-481.

Salem, G. (1986). Gender equity and the urban environment. *Urban Resources, 3,* 3-7.

Shaughnessy, R. (1986, August). Mayor Maureen. *California Journal,* pp. 395-398.

Shlay, A., & DiGregorio, D. (1985). Same city, different worlds. *Urban Affairs Quarterly, 21,* 66-86.

Sloan, B. (1983, June). Can Deep Ellum be revived? *Dallas Magazine,* pp. 20-23, 66-67.

Smith, N. (1990). *Uneven development* (2nd ed.). Cambridge, MA: Basil Blackwell.

Smith, S. (1990). Income, housing wealth and gender inequality. *Urban Studies, 27,* 67-88.

Soja, E. (1989). *Postmodern geographies.* London: Verso.

Spain, D. (1990). The effect of residential mobility and household composition on housing quality. *Urban Affairs Quarterly, 25,* 659-683.

Squires, G. (Ed.). (1989). *Unequal partnerships.* New Brunswick, NJ: Rutgers University Press.

Stone, C. (1989). *Regime politics.* Lawrence: University Press of Kansas.

Stone, C., & Sanders, H. T. (Eds.). (1987). *The politics of urban development.* Lawrence: University Press of Kansas.

Strauss, A. (1992). Community diversity. *Public Management, 74,* 22-24.

Suchman, D., & Lamb, M. (1991). West Dallas poised for change. *Urban Land, 50,* 10-16.

Swanstrom, T. (1985). *The crisis of growth politics.* Philadelphia: Temple University Press.

Tatum, H. (1985, October 11). From past to future [editorial]. *Dallas Morning News,* p. 23.

Thomas, M., & Holmes, B. (1992). Determinants of satisfaction for Blacks and Whites. *Sociological Quarterly, 33,* 459-472.

Turner, R. (1992). Growth politics and downtown development. *Urban Affairs Quarterly, 28,* 3-21.

Turner, R. (1993). Growth management decision criteria. *State and Local Government Review, 25,* 186-196.

Vogel, R. K., & Swanson, B. E. (1989). The growth machine versus the anti-growth coalition. *Urban Affairs Quarterly, 25,* 63-85.

Wilson, W. J. (1987). *The truly disadvantaged.* Chicago: University of Chicago Press.

Name Index

Subject Index

Affirmative action, 63
Affordable housing, gender and politics
of, 119-134
African American political power, rise
and decline of in Richmond,
136-153
African American women, 49, 173
as council members, 172
as heads of households, 237
as mayors, 157, 160, 172
battered, 228
domestic roles and commuting among,
244-246
earnings of, 237
economic status and commuting
among, 241-244
spousal abuse of, 220
suburban residence of, 249-250
transportation access and commuting
among, 246-248
See also Richmond, Virginia
Aid to Families With Dependent
Children (AFDC), 201, 203, 204,
206, 214, 266
clients, 209, 210, 211, 215
families receiving, 264-265
Alcoa, 69
Allegheny Conference for Community
Development, 72
American cities, gender regimes of, 44-56
American School Board Journal, 171
Apoyo, 228

Beguinages, 259-261
children in, 261
description of, 256
for controlling surplus populations,
257-258
public housing and, 256-268

Beguine leaders, 260
Beguines, 265-266, 268
as celibate, 261
Bilingual Education Parents Advisory
Committee (Chelsea, MA), 78, 90,
91
Boston, 60, 61, 70
economic restructuring and female
workers in, 62, 65-69
history of female workers in, 66
Boston School Committee, 69
Boston University, 78, 86, 87, 90, 91
School of Management of, 87
British National Conference on Preventing
Crime Against Women, 106
Brown University, 182
Brown v. Board of Education, 141
Business and Professional Women, 113

Canadian municipal initiatives, 107-111
Center for the American Woman and
Politics (CAWP), 156
Central-city development policy, gender
and, 271-288
Chelsea, Massachusetts, 78, 93, 95
Latina women's collective action in,
86-92
public school system of, 86-92
Chicago School of Sociology, 79
Child care, women's need for, 201-216
Child Care and Development Block
Grant program, 216
Child care workers, training for, 215
Cities:
as institutional intersections, 46-48
as nexus of family, economy, and
state, 45
as patriarchal, 44
gender regimes of, 44-56

About the Authors

CAROLINE ANDREW is a Professor of Political Science at the University of Ottawa, in Ottawa, Canada. Her research interests include local politics and women and politics. She is currently engaged in a research project looking at social policy at the municipal level. She is the past president of the Canadian Research Institute for the Advancement of Women (CRIAW), a nongovernmental organization that brings together community-based, university-based, and government-based researchers to promote research aimed at improving the position of women in Canadian society. She coedited *Life Spaces* (1988), a collection of texts on gender relations and urban development, with Beth Moore Milroy. Other recent publications include "Laughing Together: Women's Studies in Canada" (*International Journal of Canadian Studies*, 1990) and "The Feminist City," in *Political Arrangements: Power and the City*, edited by H. Lustiger-Thaler (Black Rose Books, 1992).

LYNN M. APPLETON is an Associate Professor in the Department of Sociology and Social Psychology at Florida Atlantic University. Much of her earlier work was on urban politics and fiscal austerity, but, increasingly, she is working on issues of gender and culture such as those in this volume's essay. Her interest in these issues was heightened by the seminal work of former colleague Martin Levine on the "gay ghetto."

SUSAN ABRAMS BECK is Associate Professor of Political Science at Fordham University—College at Lincoln Center and serves as a Deputy Editor of the *Social Science Journal*. Her recent publications include "The Fall of James Wilson's Democratic Presidency," in *The Federalists, the Antifederalists, and the American Republican Tradition*, edited by Wilson Carey McWilliams and Michael T. Gibbons (Greenwood, 1992), and "Rethinking Local Governance: Gender Distinctions on Municipal Councils," in *Gender and Policymaking: Studies of Women in Office*, edited by Debra Dodson (Center for the American Woman and Politics, 1991). The chapter in this volume is part of a larger study she is conducting on the impact of women in local office. She received her Ph.D. from Columbia University.

CHARLES S. BULLOCK III is the Richard B. Russell Professor of Political Science at the University of Georgia. He received his Ph.D. from Washington University in St. Louis and has done research on Congress, civil rights, and policy implementation. He is the author, coauthor, or coeditor of *Law and Social Change* (1972), *Racial Equality in America* (1975), *Coercion to Compliance* (1976), *Public Policy and Politics in America* (2nd ed., 1984), *Public Policy in the Eighties* (1983), *Implementation of Civil Rights Policy* (1984), *Government in America* (1984), *The Georgia Political Almanac* (1991, 1993), *Run-Off Elections in the United States* (1992), and *Forest Resource Policy* (1993). Bullock is past president of the Southern Political Science Association. He has served as chair of the Legislative Studies Section and on the Executive Committee of the American Political Science Association. He is on the editorial boards of the *Journal of Politics* and *Social Science Quarterly*. In 1991, he received the William A. Owens Creative Research Award, and in 1993 he was one of four Arts and Sciences faculty members recognized for outstanding teaching.

SUSAN E. CLARKE (Ph.D. University of North Carolina, 1979) is Associate Professor of Political Science and Director of the Center for Public Policy Research, University of Colorado at Boulder. She coedited and contributed to *The New Localism* and *Urban Innovation and Autonomy*; her journal articles and book chapters address local economic development, urban policy, women and local politics, and interest representation, among other topics. This research has been supported by the National Science Foundation and the U.S. Economic Development Administration. She has served in editorial positions and as editorial board member for several journals. In addition to academia, she spent 2 years as a Visiting Scholar in the Policy Development and Research Division at the U.S. Department of Housing and Urban Development. In 1991-1992, she served as president of the APSA Urban Politics Section. During 1993-1994, she is a Visiting Fellow at the University of Essex.

JUDITH A. GARBER received her Ph.D. from the University of Maryland and teaches political science at the University of Alberta, where she is Assistant Professor. She is currently conducting research on the competing normative theories of American cities as expressed in law, as well as comparing (with Robyne Turner) the policy responses to economic decline in Edmonton, Alberta, and Houston, Texas. Previous research on land use regulation and local authority in U.S. cities has appeared in the *Journal of Urban Affairs*, *Urban Affairs Quarterly*, and *Urban Resources*.

LOUISE JEZIERSKI is Assistant Professor of Sociology and in the Program in Urban Studies at Brown University. She is presently writing a book, *Consent to the City*, based on her research on the politics of postindustrial transformation in Cleveland and Pittsburgh. She has also published articles on the implication of public-private partnerships for governance and on postmodern urban theory. Her current research is on identity formation and community empowerment of Latina/Latino communities in Boston, Hartford, and Providence.

IBIPO JOHNSTON-ANUMONWO is an Assistant Professor of Geography at the State University of New York College at Cortland where she teaches Human, Urban, and Economic Geography, Africa, Global Development; Geography and Gender; and Women and Work. She received her bachelor's degree at the University of Ibadan in Nigeria. She has her master's degree from the Harvard Graduate School of Education and a doctorate from Clark University, Worcester, Massachusetts. She also maintains an active interest in the status of women in less developed countries. She received awards for her studies on women's employment from the Ford Foundation Women in Development Summer Institute and the American Association of Geographers Perspectives on Women Specialty Group. Her latest research efforts (partly supported by a grant from the New York State African American Research Institute) extend previous work on gender differences to incorporate race differences in employment accessibility. Her research has been published in key geography journals, and she plans to continue interdisciplinary research on the connections between home and work with a focus on gender roles and economic adjustment in urban labor markets.

LYN KATHLENE, Ph.D., is Assistant Professor of Political Science at Purdue University. Her research revolves around democratization of the policy process, which she pursues through her interest in gender politics, feminist methodology, and citizen participation. Her recent work has appeared in the *American Political Science Review, Journal of Policy Analysis and Management, Western Political Quarterly, Women's Review of Books*, and *Chronicle of Higher Education*, as well as chapters in several edited books. She is currently working on a book manuscript entitled "Gender, Public Policy, and the Legislative Process," based on research from an 18-month study funded by the Center for the American Woman and Politics, Eagleton Institute of Politics, Rutgers University. Two of her conference papers on women and politics have won regional professional association awards. Since arriving at Purdue, she has been actively working with a group of students studying the problem of rape on campus. In 1991, her rape research received a commendation from the Council on the Status of Women at Purdue.

SUSAN A. MacMANUS is Professor of Public Administration and Political Science at the University of South Florida, Tampa. She has published numerous articles on female representation in the *Journal of Politics, American Politics Quarterly, Urban Affairs Quarterly, Social Science Quarterly, National Civic Review*, and *Municipal Year Book 1993*, among others. She is the author/editor of *Reapportionment and Representation in Florida: An Historical Collection*. She also frequently serves as a consultant to state and local governments across the United States on electoral issues.

SARA McLAFFERTY is Associate Professor of Geography at Hunter College of the City University of New York. She holds M.A. and Ph.D. degrees in geography from the University of Iowa. She is the coauthor, with Avijit Ghosh, of *Location Strategies for Retail and Service Firms* (D. C. Heath), and has published widely in the fields of urban, economic, and medical geography. Her research examines questions of geographical accessibility to employment, health care services, and retail activities for women and minorities. With Valerie Preston, she is currently analyzing minority women's commuting patterns and spatial access to employment in the New York metropolitan region.

KRISTINE B. MIRANNE is a recent Ph.D. graduate of the College of Urban and Public Affairs, University of New Orleans. Her research and teaching interests include several major areas of inquiry: the social organization of housing in relation to elderly women; the analysis of the relationship between spatial institutions, gender status, and knowledge; and how the issues of aging, poverty, and gender are linked to housing policy.

MERRY MORASH received her Ph.D. in criminal justice and criminology from the University of Maryland and also holds an M.S.W. She was a practicing social worker for several years, working in such settings as a medium-security prison, juvenile court, and a neighborhood settlement house. Currently, she is the Director of the School of Criminal Justice at Michigan State University. She is actively involved in the Women's Studies Program and has just ended a 2-year term as chair of the Women's Advisory Committee to the Provost, which focuses on improving the climate for women students and faculty at Michigan State. Her research interests include delinquency causation and control. She is the author of the textbook, *Juvenile Delinquency: Concepts and Control.* Her other research has focused on women as offenders and victims.

VALERIE PRESTON is an Associate Professor of Geography at York University, Canada. Trained as a social geographer, her current research examines spatial aspects of women's work. She is involved in a collaborative project that uses secondary data to examine how race and gender influence commuting in New York. With an anthropologist, she is exploring the changing spatial and sectoral patterns of immigrant women's paid employment in Toronto and the nature of home work done by immigrant women in Toronto. The studies are intended to demonstrate how economic restructuring has altered the relationships between women's paid and unpaid work at the workplace and in the home.

GORDANA RABRENOVIC is Assistant Professor of Sociology at Northeastern University at Boston. Her research and teaching interests are oriented toward comparative, cross-cultural social analysis in the areas of urban sociology, social movements, and voluntary and nonprofit organizations. She is currently working on a book that compares neighborhood mobilization in two different type of cities: declining manufacturing cities and service sector cities.

LEWIS A. RANDOLPH is Assistant Professor in the Political Science Department/Public Administration Program at Ohio University. He received his M.A. in political science at the University of Illinois at Champaign/Urbana and his Ph.D. in political science at Ohio State University. His areas of interest are urban development, urban politics, public policy, and Black politics. His most recent publication, coauthored with William W. Goldsmith, is "Ghetto Economic Development," in

Theories of Local Economic Development, edited by Richard Bingham and Robert Meir.

ANNA M. SANTIAGO received her Ph.D. in urban social institutions from the University of Wisconsin—Milwaukee. She is currently on the faculty in the Department of Sociology at Indiana University where she also is affiliated with the Population Institute and the Women's Studies Program. Her research interests focus on interethnic patterns of residential segregation in the United States, long-term poverty and welfare dependency, urban economic development, and the economic consequences of disability status on minority populations and Latino families. Her interest in families and social service use by minorities stems from several years' work as a social worker. Her research in these areas has appeared in journals such as *Social Problems, Urban Affairs Quarterly, Urban Geography*, and *Hispanic Journal of Behavioral Sciences*.

DAPHNE SPAIN holds a Ph.D. in sociology from the University of Massachusetts and is a faculty member in the Department of Urban and Environmental Planning in the School of Architecture at the University of Virginia, where she is currently serving as Acting Dean. She has published in the *Journal of the American Planning Association*, the *Journal of Planning Education and Research*, and *Sociological Theory*; her most recent book is *Gendered Spaces* (University of North Carolina Press, 1992).

LYNN A. STAEHELI is an Assistant Professor in the Department of Geography and a Faculty Research Associate in the Institute of Behavioral Science at the University of Colorado, Boulder. Her research interests include gender and race relations in urban politics.

GAYLE T. TATE is Assistant Professor of Political Science in Africana Studies at Rutgers University. She has published articles in such varied journals as *Women & Politics, National Political Science Review*, and *Western Journal of Black Studies*. She was a 1993 Rockefeller Humanities Fellow at the City University of New York. Her research interests focus on the political consciousness of Black women and Black political mobilization.

ROBYNE S. TURNER is Associate Professor of Political Science at Florida Atlantic University. She received her Ph.D. in political science from the University of Florida. Her publications have appeared in *Urban Affairs Quarterly*, *Journal of Urban Affairs*, *Publius: The Journal of Federalism*, and *State and Local Government Review*. She is writing a book on the politics of downtown development in Sunbelt cities, and her other areas of research include community development initiatives and sustainable community development for women. She is currently participating in an evaluation of a national demonstration project to develop community development corporations.

ALMA H. YOUNG is Professor of Urban Studies and Planning and Director of the Ph.D. program in Urban Studies, College of Urban and Public Affairs at the University of New Orleans. She has been actively involved in civic activities involving women and children in the City of New Orleans. Her research areas include the political economy of urban development, Third World women and development, and issues of gender within the urban environment.